AF361741

Fracture, Fatigue, and Structural Integrity of Metallic Materials and Components Undergoing Random or Variable Amplitude Loadings

Fracture, Fatigue, and Structural Integrity of Metallic Materials and Components Undergoing Random or Variable Amplitude Loadings

Editors

Denis Benasciutti
Mark T. Whittaker
Turan Dirlik

MDPI • Basel • Beijing • Wuhan • Barcelona • Belgrade • Manchester • Tokyo • Cluj • Tianjin

Editors

Denis Benasciutti
Department of Engineering
University of Ferrara
Ferrara
Italy

Mark T. Whittaker
Institute of Structural
Materials
Swansea University
Swansea
United Kingdom

Turan Dirlik
Dirlik Controls Ltd
Dirlik Controls Ltd
Rugby
United Kingdom

Editorial Office
MDPI
St. Alban-Anlage 66
4052 Basel, Switzerland

This is a reprint of articles from the Special Issue published online in the open access journal *Metals* (ISSN 2075-4701) (available at: www.mdpi.com/journal/metals/special_issues/fracture_fatigue_structural_integrity).

For citation purposes, cite each article independently as indicated on the article page online and as indicated below:

LastName, A.A.; LastName, B.B.; LastName, C.C. Article Title. *Journal Name* **Year**, *Volume Number*, Page Range.

ISBN 978-3-0365-4692-6 (Hbk)
ISBN 978-3-0365-4691-9 (PDF)

Contents

About the Editors

Denis Benasciutti

Denis Benasciutti is Associate Professor of Machine Design at the University of Ferrara, Italy. In 2001 he graduated cum laude in Materials Engineering at the University of Ferrara, where, in 2005, he also received his Ph.D. In 2006, he earned a Master in Welding Engineering and the diploma as European/International Welding Engineer (EWE/IWE). From 2006 to 2015, he was Assistant Professor at the University of Udine, Italy. His research interests involve the structural integrity of mechanical components subjected to stochastic uniaxial and multiaxial loadings (with special focus on frequency-domain spectral methods) and the cyclic plasticity and low-cycle fatigue behaviour of traditional and additive manufactured materials. He published over 80 scientific articles in peer reviewed journals, on the topics of metal fatigue and structural durability, finite element method, "energy harvesting", cyclic plasticity of metals.

Mark T. Whittaker

Professor Mark Whittaker is the Director of the Institute of Structural Materials at Swansea University, and has a background in high-temperature materials. Following his Ph.D. in 2003, he undertook postdoctoral research mainly based around fatigue lifing of titanium alloys, before accepting an RCUK Research Fellowship in 2007, leading to promotion to Senior Lecturer in 2012, Associate Professor in 2015, and Professor in 2019.

His research interests over this time have extended to high temperature lifing of materials with a particular focus on material behaviour under creep and thermo-mechanical fatigue conditions. Collaboratively working with Professor Brian Wilshire for 5 years enabled the further development of the Wilshire Equations, with the aim to relate concepts from the equation, such as activation energy to micromechanical behaviour, an activity which is still being pursued currently.

Prof Whittaker has worked extensively with industrial partners including Rolls-Royce, TIMET, Pratt & Witney, and Forged Solutions. He also collaborates with national and international research partners and has published over 60 papers in peer reviewed journals, also making invited presentations at a number of international conferences.

Turan Dirlik

Turan Dirlik is the managing director of Dirlik Controls Ltd, UK, supplying bespoke materials testing software to research institutions. Turan's Ph.D. thesis focused on developing a fatigue life assessment technique, using the rainflow cycle counting process. This method is now commonly known as the *Dirlik Formula*. His current work includes fatigue crack propagation and thermo-mechanical fatigue.

Editorial

Fracture, Fatigue, and Structural Integrity of Metallic Materials and Components Undergoing Random or Variable Amplitude Loadings

Denis Benasciutti [1,*], Mark T. Whittaker [2] and Turan Dirlik [3]

[1] Department of Engineering, University of Ferrara, Via Saragat 1, 44122 Ferrara, Italy
[2] Institute of Structural Materials, Swansea University, Crymlyn Burrows, Swansea SA1 8EN, UK; m.t.whittaker@swansea.ac.uk
[3] Dirlik Controls Ltd., 12 Main Street, Rugby CV22 7NB, UK; turan@dirlik.co.uk
*   Correspondence: denis.benasciutti@unife.it

Citation: Benasciutti, D.; Whittaker, M.T.; Dirlik, T. Fracture, Fatigue, and Structural Integrity of Metallic Materials and Components Undergoing Random or Variable Amplitude Loadings. *Metals* **2022**, *12*, 919. https://doi.org/10.3390/met12060919

Received: 13 May 2022
Accepted: 23 May 2022
Published: 27 May 2022

Publisher's Note: MDPI stays neutral with regard to jurisdictional claims in published maps and institutional affiliations.

1. Introduction and Scope

When quickly reviewing engineering and industrial fields, one often discovers that a large number of metallic components and structures are subjected, in service, to random or variable amplitude loadings. The examples are many: vehicles subjected to loadings and vibrations caused by road irregularity and engine, structures exposed to wind, off-shore platforms undergoing wave-loadings, and so on.

Just like constant amplitude loadings, random and variable amplitude loadings can make fatigue cracks initiate and propagate, even up to catastrophic failures. Engineers faced with the problem of estimating the structural integrity and the fatigue strength of metallic structures, or their propensity to fracture, usually make use of theoretical or experimental approaches, or both. Counting methods (e.g., rainflow) provide information on the fatigue cycles in the load, whereas damage accumulation laws (as the celebrated Palmgren–Miner linear rule) establish how to sum up the damage of each counted cycle. In structural integrity, this is named as the "time-domain" approach. Over recent years, the "frequency-domain" approach has also received increasing and widespread use, especially with random loadings; this approach estimates fatigue life based on load statistical properties represented, in the frequency domain, by a power spectral density. Neither of the previous approaches, however, can do without the support of experimental laboratory testing, which provides a means to collect material strength data under specific loading conditions, or to verify preliminary estimations.

The purpose of this Special Issue is to collect articles aimed at providing an up-to-date overview of approaches and case studies—theoretical, numerical or experimental—on several topics in the field of fracture, fatigue strength, and the structural integrity of metallic components subjected to random or variable amplitude loadings.

2. Contributions

The Special Issue counts a total of ten articles on a variety of topics, including fracture mechanics, experimental measurement and testing, and structural integrity assessment.

Šmíd et al. [1] investigate the role of microstructure on the high-temperature fatigue strength of a polycrystalline Ni-based superalloy used for manufacturing gas turbine blades by investment casting. Alloys made by three different casting setups, with or without the application of hot hydrostatic pressing (HIP), are examined to identify the inter-relationship between the casting process, microstructure, and fatigue life. The experimental findings suggest that the fatigue life of the tested superalloy is mainly related to the porosity retained after casting and to the effect of HIP treatment, with a prominent role of porosity size; by contrast, grain size and texture have a minor effect and contribute mostly to fatigue life scatter.

Uzun and Korsunsky [2] apply the height digital image correlation (hDIC) technique for determining the discontinuous displacement field associated with cracks on the specimen or component surface. Their paper implements graphics processing unit (GPU) parallel computing to improve the computational performance of the hDIC technique when it is applied to high-density topography data necessary to determine discontinuous displacements. The proposed hDIC technique with GPU parallel computation method is successful in identifying discontinuous edges in a 6082/HE30 aluminum alloy specimen subjected to tensile testing until failure.

Three papers [3–5] focus on fracture mechanics analysis under spectrum loadings. Jones et al. [3] present a fracture mechanics approach in the context of damage tolerant design and economic service life assessment of military helicopters—where "economic life" indicates the period during which repairing a component is more cost-effective than replacing it. The results of a computational study reveal that the Hartman–Schijve variant of the NASGRO crack growth equation—extensively validated against long crack growth data—can also model the small crack growth in the AA7075-T7351 alloy under a simplified or reduced helicopter spectrum. The role of statistical variability in crack growth histories is also pointed out.

Related to small-crack growth under spectrum loading is also the article by Newman [4], which deals with the "rainflow-on-the-fly" subroutine implemented in FASTRAN, an advanced nonlinear crack closure-based numerical code for fatigue life prediction. Laboratory fatigue tests conducted on compact and single-edge-notch-bend specimens made of a 9310 steel are used to validate the "rainflow-on-the-fly" subroutine under several specially designed variable-amplitude loading sequences. Results indicate that the damage is a function of load history and that the usual rainflow counting method would not be able to capture the crack-growth damage correctly, unless the method is updated during crack-growth history.

Boljanović and Carpinteri [5] develop a novel computation framework to investigate the effect of overloads (e.g., retardation effect) on the crack growth and fatigue life of cracked plates. The study considers the Huang–Moan crack growth model to account for R-ratio effect, and includes crack retardation through Wheeler's model. As a case study, the paper analyzes a 6061-T6 aluminum plate with edge crack subjected to several types of constant amplitude loading characterized by different R and overload ratios; besides axial loading, inclined loading corresponding to mixed I+II mode loading is also considered. The paper concludes that a higher overload ratio induces more pronounced retardation effects, with an increase in fatigue life. Other conclusions are that retardation effects due to overloading also depend upon overload crack length and loading direction.

Another study related to fracture mechanics is that by Yang et al. [6]. It applies an elastoplastic fracture analysis by the extended finite element method (XFEM) to study how thermal shock in repair welding affects the cracking behavior in the heat affected zone (HAZ) of a welded joint. A two-dimensional model with initial crack is used for simulating the thermo-mechanical shock during repair welding of a butt weld in P91 steel. In XFEM simulation, the nonlinear damage evolution (elastoplastic fracture) is represented by a cohesive crack model. This model allows for a comprehensive understanding of the evolution of both thermo-mechanical stress and damage in repair welding, and how they relate to crack propagation.

The topic of structural integrity assessment of engineering components is discussed in four papers [7–10]. Among them, the literature review of Marques et al. [7] surveys the various systems for high-cycle fatigue testing with uniaxial and multiaxial random loadings. Testing systems are compared in terms of several characteristics, as for example type of machine (servo-hydraulic, shakers), specimen geometry, relationship between the number of loading inputs applied to the specimen and the resulting local state of stress. A classification is proposed to allow the analogies, differences, advantages and possible limitations of every testing system to be emphasized.

Another review article is the work by Dirlik and Benasciutti [8]. Not only does it provide a general introduction to the frequency domain approach (spectral methods) used for assessing fatigue life directly from the power spectral density of the random loading, but it also scrutinizes in more detail two methods—Dirlik method and TB method–that are considered the most accurate by many scholars. Other than explaining the historical background that led to the development of these two methods, the paper also emphasizes their better performance by summarizing the outcomes of numerical and experimental comparative studies from the literature. It finally concludes with some recommendations for the use of spectral methods.

The paper by Marques et al. [9] is, instead, focused on evaluating the sampling variability of the fatigue damage computed in one single nonstationary time history. Damage variability is tackled by means of confidence intervals, which are constructed from the damage values of disjoint segments and blocks into which the nonstationary time history is subdivided. As a case study, the paper analyzes the strain records measured in a wheel of a telescopic handler industrial vehicle. Interesting is also a preliminary screening in which strain records are examined by two methods (short-time Fourier transform, run test) to verify whether they fulfill the hypotheses of the proposed confidence interval.

The interesting study of Pisani et al. [10] deals with the structural integrity and fatigue life estimation of two types of lattice structures in AlSi10Mg alloy (namely, face-centered cubic and diamond lattice-based), fabricated by laser powder bed fusion (L-PBF) and tested under random fatigue loadings. The structural integrity of lattice structures is estimated by a finite element model, validated on sweep dynamic tests and employed to determine the stress levels within the lattice structures when subjected to the input accelerations applied in experiments. Fatigue failure is identified by acceleration change or frequency drop in the structure. Predicted and experimental results are shown to agree quite well.

3. Conclusions and Outlook

With its ten published papers, this Special Issue covers various topics related to the fatigue strength and structural integrity assessment of engineering components subjected to variable amplitude or random loadings. Theoretical, numerical, and experimental approaches are presented, both as more methodological analyses and as case studies. We believe, as Guest Editors, that the quality and variety of the collected articles can contribute to the advancement of future research in this highly challenging research field.

Funding: This research did not receive any external funding.

Conflicts of Interest: The authors declare no conflict of interest.

References

1. Šmíd, M.; Horník, V.; Kunz, L.; Hrbáček, K.; Hutař, P. High Cycle Fatigue Data Transferability of MAR-M 247 Superalloy from Separately Cast Specimens to Real Gas Turbine Blade. *Metals* **2020**, *10*, 1460. [CrossRef]
2. Uzun, F.; Korsunsky, A. The Use of Surface Topography for the Identification of Discontinuous Displacements Due to Cracks. *Metals* **2020**, *10*, 1037. [CrossRef]
3. Jones, R.; Peng, D.; Raman, R.; Huang, P. Computing the Growth of Small Cracks in the Assist Round Robin Helicopter Challenge. *Metals* **2020**, *10*, 944. [CrossRef]
4. Newman, J. Fatigue and Crack Growth under Constant- and Variable-Amplitude Loading in 9310 Steel Using "Rainflow-on-the-Fly" Methodology. *Metals* **2021**, *11*, 807. [CrossRef]
5. Boljanović, S.; Carpinteri, A. Computational Failure Analysis under Overloading. *Metals* **2021**, *11*, 1509. [CrossRef]
6. Yang, K.; Zhang, Y.; Zhao, J. Elastoplastic Fracture Analysis of the P91 Steel Welded Joint under Repair Welding Thermal Shock Based on XFEM. *Metals* **2020**, *10*, 1285. [CrossRef]
7. Marques, J.; Benasciutti, D.; Niesłony, A.; Slavič, J. An Overview of Fatigue Testing Systems for Metals under Uniaxial and Multiaxial Random Loadings. *Metals* **2021**, *11*, 447. [CrossRef]
8. Dirlik, T.; Benasciutti, D. Dirlik and Tovo-Benasciutti Spectral Methods in Vibration Fatigue: A Review with a Historical Perspective. *Metals* **2021**, *11*, 1333. [CrossRef]

9.   Marques, J.; Solazzi, L.; Benasciutti, D. Fatigue Analysis of Nonstationary Random Loadings Measured in an Industrial Vehicle Wheel: Uncertainty of Fatigue Damage. *Metals* **2022**, *12*, 616. [CrossRef]
10.   Pisati, M.; Corneo, M.; Beretta, S.; Riva, E.; Braghin, F.; Foletti, S. Numerical and Experimental Investigation of Cumulative Fatigue Damage under Random Dynamic Cyclic Loads of Lattice Structures Manufactured by Laser Powder Bed Fusion. *Metals* **2021**, *11*, 1395. [CrossRef]

Article

Computing the Growth of Small Cracks in the Assist Round Robin Helicopter Challenge

Rhys Jones [1,2,*] , **Daren Peng** [1] , **R.K. Singh Raman** [1,3] and **Pu Huang** [1]

1   Centre of Expertise for Structural Mechanics, Department of Mechanical and Aerospace Engineering, Monash University, Clayton 3800, Victoria, Australia; daren.peng@monash.edu (D.P.); raman.singh@monash.edu (R.K.S.R.); pu.huang@monash.edu (P.H.)
2   Titomic Limited, Building 3/270 Ferntree Gully Rd, Notting Hill 3130, Victoria, Australia
3   Department of Chemical Engineering, Monash University, Clayton 3800, Victoria, Australia
*   Correspondence: rhys.jones@monash.edu; Tel.: +61-487753232

Received: 17 June 2020; Accepted: 8 July 2020; Published: 14 July 2020

Abstract: Sustainment issues associated with military helicopters have drawn attention to the growth of small cracks under a helicopter flight load spectrum. One particular issue is how to simplify (reduce) a measured spectrum to reduce the time and complexity of full-scale helicopter fatigue tests. Given the costs and the time scales associated with performing tests, a means of computationally assessing the effect of a reduced spectrum is desirable. Unfortunately, whilst there have been a number of studies into how to perform a damage tolerant assessment of helicopter structural parts there is currently no equivalent study into how to perform the durability analysis needed to determine the economic life of a helicopter component. To this end, the present paper describes a computational study into small crack growth in AA7075-T7351 under several (reduced) helicopter flight load spectra. This study reveals that the Hartman-Schijve (HS) variant of the NASGRO crack growth equation can reasonably accurately compute the growth of small naturally occurring cracks in AA7075-T7351 under several simplified variants of a measured Black Hawk flight load spectra.

Keywords: small cracks; helicopter flight load spectra; FALSTAFF flight load spectra; fatigue crack growth

---

1. Introduction

It is now known that "ab initio" design and aircraft sustainment [1,2] are best tackled using different computational tools. United States Air Force (USAF) airworthiness standard MIL-STD-1530D [3] states that analysis is the key to both damage tolerant design and to assessing the economic life of military aircraft. MIL-STD-1530D also states that the primary role of testing is "to validate or correct analysis methods and results and to demonstrate that requirements are achieved". The USAF-McDonnell Douglas study into the economic life of USAF F-15 aircraft [1] was arguably the first to reveal that sustainment analyses need to use the short crack da/dN versus ΔK curve, and not the da/dN versus ΔK curve determined as per the US American Society for Testing and Materials (ASTM) fatigue test standard ASTM E647-13a [4]. (The term durability is defined in MIL-STD-1530D [3] as: "Durability is the attribute of an aircraft structure that permits it to resist cracking, corrosion, thermal degradation, delamination, wear, and the effects of foreign object damage for a prescribed period of time". MIL-STD-1530D [3] defines the term economic service life: The economic service life is the period during which it is more cost-effective to maintain, repair, and modify an aircraft component or aircraft than to replace it.) This conclusion is now echoed in ASTM E647-13a, Appendix X3. Whereas the ability of various crack growth equations to capture the growth of long cracks under a representative helicopter flight load spectrum has been studied [5–7] as part of the "Helicopter Damage Tolerance Round-Robin" challenge [8], there are few studies into the ability of crack growth equations to model small crack growth of small under helicopter flight load spectra. This shortcoming is particularly important since the durability/economic "initial flaw assumptions" contained in the US Joint Services Structures Guidelines JSSG2006 [9], which in MIL-STD-1530D are termed as equivalent initial damage sizes (EIDS), are typically 0.05 inches (0.127 mm). Indeed, this size of EIDS is also referenced in Structures Bulletin EZ-19-01, which presents the USAF approach to the Durability and Damage Tolerance Certification for Additive Manufacturing of Aircraft Structural Metallic Parts [10]. The importance of a validated predictive capability is highlighted in MIL-STD-1530D Sections 5.2.5 and 5.2.6, which state that a damage tolerance and a durability analysis must be performed for all aircraft, and in Section 5.3.4 which states that the purpose of full scale fatigue testing is to "validate or correct the analysis". MIL-STD-1530D also states that a factor of 2 is to be used on these analyses. The Australian Defence Science and Technology (DST) Group's small crack Helicopter Round Robin Challenge [11,12] is, to the best of the author's knowledge, the first attempt to address this shortcoming, i.e., the lack of a validated analysis for small crack growth under a helicopter flight load spectrum.

It has previously been shown [7] that the Hartman-Schijve (HS) crack growth equation [2], which is an extension of a concept first proposed in [13], accurately predicted crack growth in the "Helicopter Damage Tolerance Round-Robin" challenge [8]. It has also been shown [2,7,14–18] that this formulation can also predict small crack growth under maritime aircraft, combat aircraft, and civil aircraft flight load spectra, and that the small crack equation needed for a durability/economic life analysis can often be determined from the associated long crack equation by setting the fatigue threshold term to a small value. The HS equation has also been shown to hold for crack growth in adhesively-bonded joints, bonded wood structures, and both bridge and rail steels [19–25], as well as for delamination crack growth in composite structures [26–32]. Consequently, the focus of the present paper is to examine if the HS crack growth equation can also capture crack growth in the DST Advancing Structural Simulation to drive Innovative Sustainment Technologies (ASSIST) small crack Helicopter Round Robin Challenge.

The general form of the HS equation used in this paper is as given in [2], viz:

$$\mathrm{d}a/\mathrm{d}N = D \, (\Delta\kappa)^n \tag{1}$$

where a is the crack length/depth, N is the number of cycles, D is a material constant and n is another material constant that is often approximately 2. The crack driving force $\Delta\kappa$ used in Equation (1) was first suggested by Schwalbe [33], viz:

$$\Delta\kappa = (\Delta K - \Delta K_{\mathrm{thr}})/(1 - K_{\mathrm{max}}/A)^{1/2} \tag{2}$$

here K is the stress intensity factor, K_{max} and K_{min} are the maximum and minimum values of the stress intensity factor seen in the cycle, $\Delta K = (K_{\mathrm{max}} - K_{\mathrm{min}})$ is the range of the stress intensity factor that is seen in a cycle, ΔK_{thr} is the "effective fatigue threshold", and A is the cyclic fracture toughness. As per [2,14,16,18], the values of the terms ΔK_{thr} and A are chosen to fit the measured data. As further explained in [34], the term ΔK_{thr} is related to the ASTM E647-13a definition of the fatigue threshold ΔK_{th}, namely the value of ΔK at a value of da/dN of 10^{-10} m/cycle, by:

$$\Delta K_{\mathrm{th}} = \Delta K_{\mathrm{thr}} + (10^{-10}/D)^{1/n} \tag{3}$$

as a general rule, crack growth predictions made using Equations (1) and (2) are quite sensitive to the value used for ΔK_{thr}, and relatively insensitive to the value of A.

The HS equation has also been shown [34–38] to capture the growth of both small and long cracks in additively manufactured materials (AM), and has an ability to account for the effect of residual stresses in both conventionally and additively manufactured materials [39]. It has also been shown to be able to capture the effect of surface roughness on the fatigue life of a component [39]. This finding is particularly important given the statement by the Under Secretary of Defense, Acquisition and Sustainment [40] that "AM parts can be used in both critical and non-critical applications", and the statement in the USAF Structures Bulletin EZ-19-01 [10] that for AM parts that the most difficult challenge is to establish an "accurate prediction of structural performance" specific to durability and damage tolerance (DADT). As such it is envisaged that if it can be shown that the HS equation can be shown to reasonably accurately compute the growth of small cracks subjected to helicopter flight load spectra, then it may be useful for assessing if an AM (helicopter) replacement part, or an AM repair to a helicopter part, meets the durability requirement inherent in the Structures Bulletin EZ-19-01.

2. Materials and Methods

The majority of references quoted in this paper are taken from peer reviewed journals. The refereed conferences, proceedings, and texts referenced are either publicly available, or available from Google searches. Thirty-nine of the journal papers referenced are in journals that are either listed in SCOPUS or in the World of Science (WOS). The book chapters referenced are listed in SCOPUS. In the case of conference papers, one is in the Proceedings of 13th International Conference on Fracture (ICM13), two are contained in the Proceedings of the 1st Virtual European Conference on Fracture, two are available on Research Gate; seven references are available on various US Department of Defense DTIC websites, one is available on the American Helicopter Society website, and another is on the US Pentagon website. Keywords that were used in these searchers were durability, damage tolerance, Hartman-Schijve, small cracks, additive manufacturing, crack growth in operational aircraft, and aircraft certification.

The paper begins by using the HS equation [2] to compute crack growth in an AA7075-T7352 specimen under a FALLSTAFF (which is an industry standard combat aircraft spectrum) flight load spectrum. It is then used evaluate crack growth under several variants of a US Army Blackhawk spectrum.

3. Crack Growth in 7075-T7351

3.1. Crack Growth under a FALSTAFF Flight Load Spectrum

Before we can compute crack growth in the Helicopter Challenge, we need to establish the constants in the HS equation. To do this we examined the crack length histories given in [41] for the growth of through-the-thickness cracks in a 6.35 mm thick middle tension (MT) panel, with a rectangular cross section, tested under a FALSTAFF flight load spectrum. A plan view of the specimen geometry is shown in Figure 1. The specimens were pre-cracked to a length of approximately 2 mm before the main fatigue test. The specimens were then tested under FALSTAF, an industry-standard combat aircraft load spectrum. The test loads were applied at a frequency of 10 Hz, see [41]. The maximum load in the spectrum was 60 kN. This corresponds to a remote stress of 157.48 MPa in the working section. One block of FALSTAFF load spectrum consisted of 9006 cycles. This equates to 100 equivalent flight hours. The various crack growth histories for the 25 tests performed in [41] are shown in Figure 2.

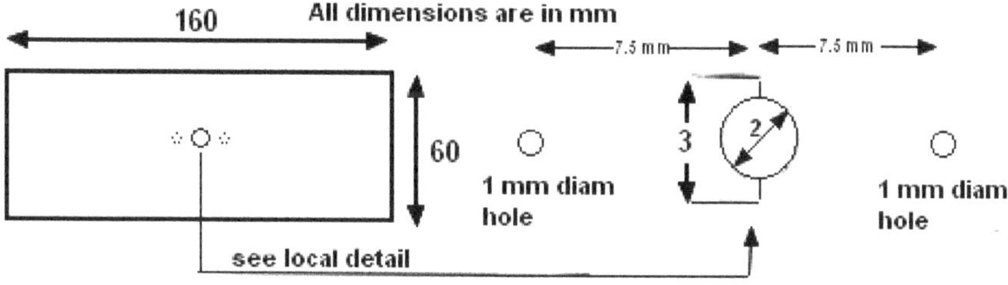

Figure 1. Schematic diagram of the MT specimen used.

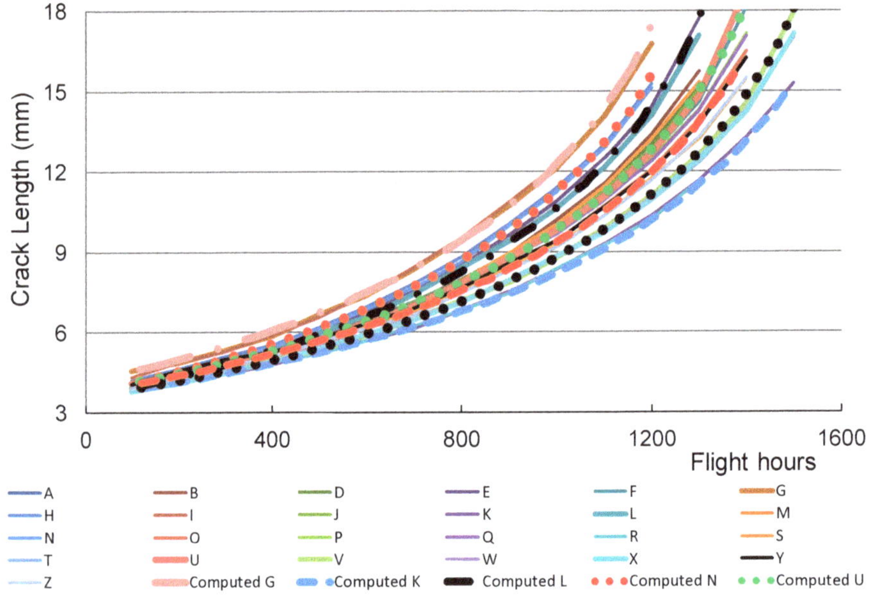

Figure 2. Plot showing the cycle-by-cycle analysis of MT specimens under FALSTAFF load spectrum.

The similarity between the da/dN versus ΔK crack growth curves associated with AA7075-T6 and AA7075-T7351 meant that the values of the constants D and n in Equation (1) for AA7075-T7351 could be taken to be as given in [14] for AA7075-T6, namely: $D = 1.86 \times 10^{-9}$ (MPa^{-2} cycle^{-1}), and $n = 2$. The value of A was taken from that given in [14] for tests on small cracks in AA7075-T7351, viz: $A = 111$ MPa $\sqrt{\text{m}}$. A similar value is given in [42]. The resultant measured and computed crack growth histories are shown in Figure 2, and the values of A and ΔK_{thr} used in the analysis are given in Table 1. Figure 2 reveals excellent agreement between the measured and computed crack growth histories. Figure 2 also reveals that, as reported in [2,15,16,18,38,43], the scatter in the crack growth histories can be captured by allowing for variability in the term ΔK_{thr}. Figure 3 presents the crack growth history plotted using log-linear axes. Figure 3 reveals that crack growth in these 25 tests could be approximated as being exponential. As explained in [2] this is due to the test program being performed on cracks in an MT panel.

Table 1. Values of A and ΔK_{thr} used in Figure 2 when computing the crack growth curves for the various tests.

Test	A (MPa $\sqrt{\text{m}}$)	ΔK_{thr} (MPa $\sqrt{\text{m}}$)
G	111	1.3
K	111	1.79
L	111	1.1
N	111	1.1
U	111	1.48
V	111	1.6
X	111	1.45
Y	111	1.6
G	111	1.3

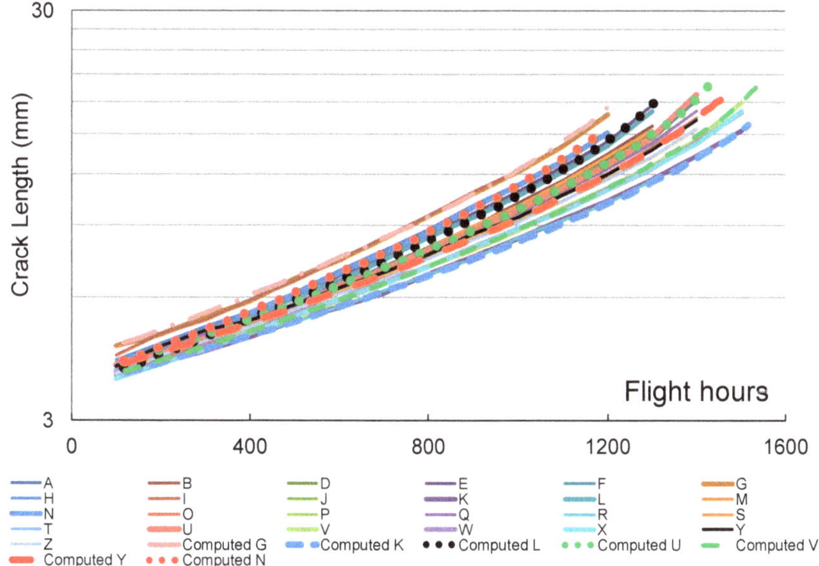

Figure 3. The crack growth history plotted using log-linear axes.

3.2. Short Cracks in 7075-T7351

Having determined the crack growth equation for AA7075-T7351 a comparison between the $R = 0.8$ AA7075-T73 short crack da/dN versus ΔK curve presented in [44] and the corresponding curve predicted using Equation (1) with $D = 1.86 \times 10^{-9}$, and $n = 2$, $\Delta K_{\text{thr}} = 0.6$ MPa $\sqrt{\text{m}}$ and $A = 111$ MPa $\sqrt{\text{m}}$ is given in Figure 4. Figure 4 reveals that there is an excellent agreement between the computed and the measured curve presented in [44]. The next section will use these values of D, n, ΔK_{thr}, and A to compute crack growth in the DST Helicopter Challenge.

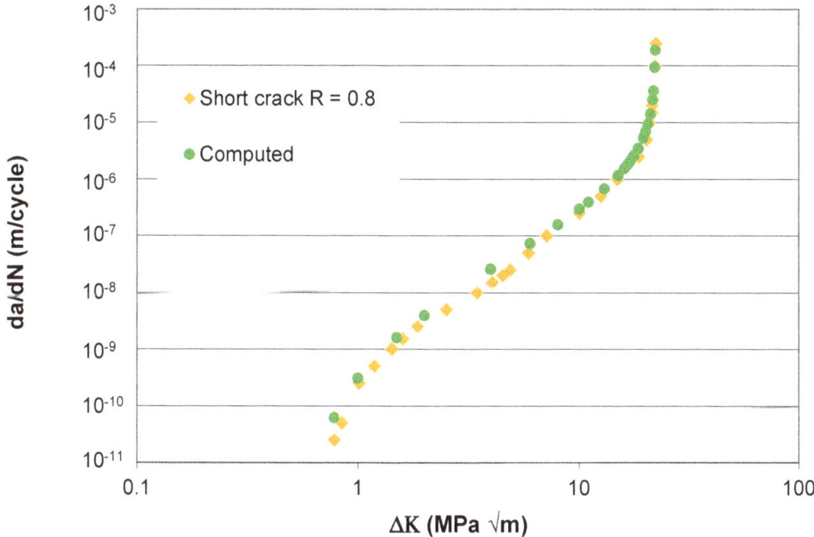

Figure 4. Comparison with the small/short crack growth curve given in [44].

4. Computing Crack Growth in the DST Small Crack Helicopter Round Robin Challenge

The focus of problem proposed in the (DST) Group's small crack Helicopter Round Robin Challenge was to compute the growth of small cracks in 8.4 mm thick AA7075-T7351 specimens under a range of simplified helicopter flight load spectra [11,12]. The baseline spectrum, which is described in [45], was obtained from a flight strain survey conducted on a US Army H-60 Black Hawk helicopter. The crack growth data and details of the specimen and the various helicopter flight load spectra were made publicly available via the DST ASSIST initiative and are available at [11,12].

The load sequences provided by DST as part of the ASSIST Round Robin were termed IRF-E14, IRF-E15, and IRF-E16. These spectra are simplified/reduced versions of the baseline spectrum, where different numbers of small amplitude cycles have been removed. Sequences termed CSL090SSXX, which are truncated versions of the IRF-E16 spectrum, were also provided. The CSL090SSXX spectra had: (a) an additional 90% of the smallest cycles removed, and (b) the mid-range cycles were scaled by XX%.

A plan view of the test specimens used by DST in the ASSIST test program [11] is shown in Figure 5. The number of turning points in each of the spectra used in this test study are given in Table 2. The surface of the specimen was etched to promote organic crack nucleation, using a solution of Hydrofluoric acid (1%), Nitric acid (50%), and water (49%). Further details of the test specimen and the spectra are given in [11,12].

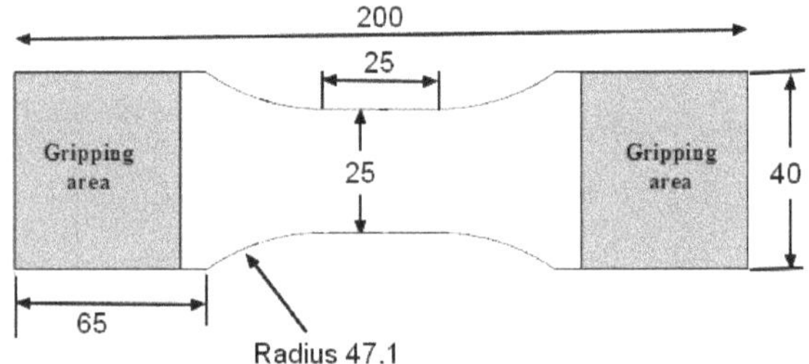

Figure 5. Geometry of the 8.4 mm thick test specimen. (units are in mm).

Table 2. The number of turning points in each spectrum.

IRF-E15	IRF-E15	IRF-E16	CSL090SS00	CSL090SS05	CSL090SS15	CSL090SS20
82,839	248,255	649,666	64,958	64,958	64,958	64,958

Short Cracks in 7075-T7351

Equation (1), with $D = 1.86 \times 10^{-9}$, $n = 2$, and $\Delta K_{thr} = 0.6$ MPa $\sqrt{m}$, was used to predict the crack growth histories associated with the Round Robin tests subjected to the following spectra: IRF-E14, IRF-E15, IRF-E16, CSL090SS00, CSL090SS05, CSL090SS15, and CSL090SS20. The analysis was performed using both $A = 40$ MPa $\sqrt{m}$, and $A = 111$ MPa $\sqrt{m}$. The value of $A = 40$ MPa $\sqrt{m}$ was investigated since prior DST constant amplitude tests [41] had yielded values of $A \approx 32$ MPa $\sqrt{m}$ for twenty four mm thick AA7075-T7351 specimens, and $A \approx 40$ MPa $\sqrt{m}$ for three mm thick AA7075-T7351 specimens. The value of $A = 111$ MPa $\sqrt{m}$ was investigated since it is associated with the short crack tests reported in [44]. As per the requirements enunciated in the ASSIST challenge [11], the initial crack size was taken to be a centrally located 0.01 mm semi-circular surface crack. The stress intensity factors were computed using the methodology outlined in [46]. Comparisons between the measured and computed crack growth histories are given in Figures 6–12, where the computed crack depth histories are labelled "Computed $\Delta K_{thr} = 0.6\ A = XX$", where XX is either 40 or 111 depending on what value of A was used in the analysis. Here it should be noted that, as shown in Figures 6–12, each spectrum test program had a number of repeated tests. Figures 6–12 reveal that there is little difference between the crack growth histories computed using $A = 40$ MPa $\sqrt{m}$ or $A = 111$ MPa $\sqrt{m}$. This is because the majority of the life is consumed in growing to a depth of 1 mm. We also see that there is reasonable agreement between the measured and predicted crack growth curves.

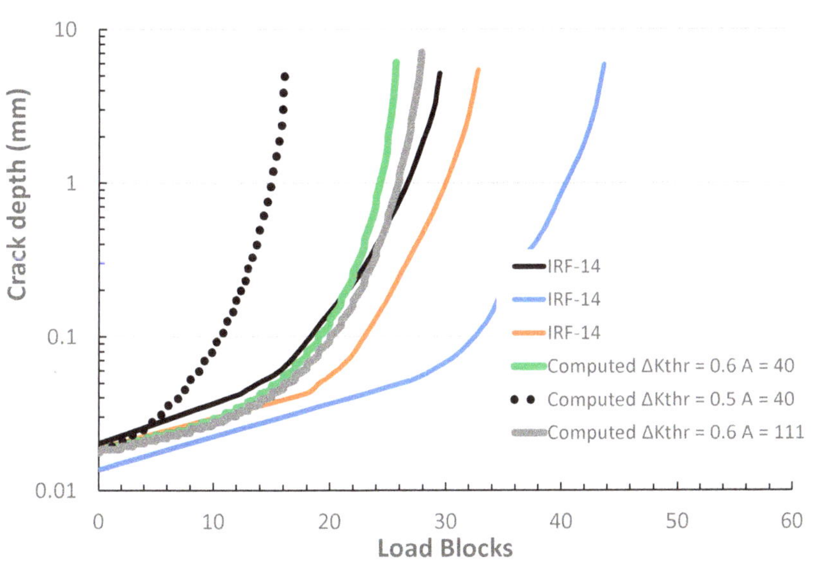

Figure 6. The measured and computed crack depth histories for the helicopter flight load spectrum IRF-E14.

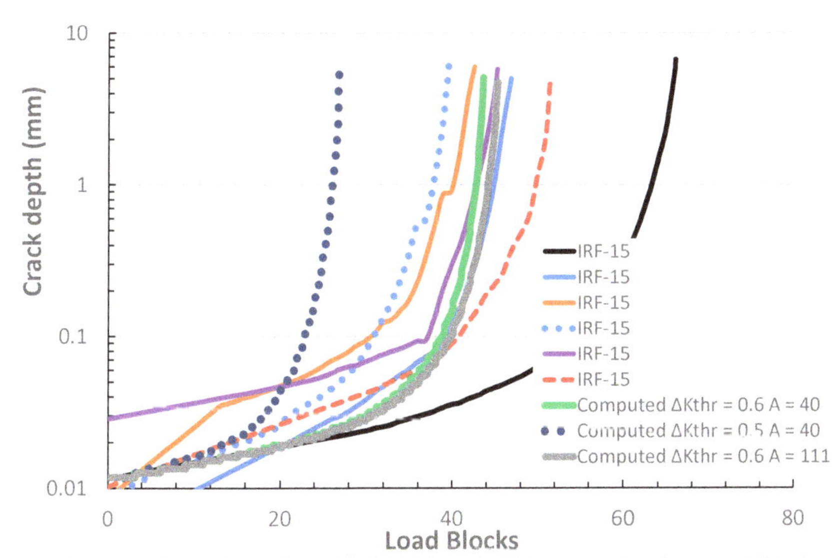

Figure 7. The measured and computed crack depth histories for the helicopter flight load spectrum IRF-E15.

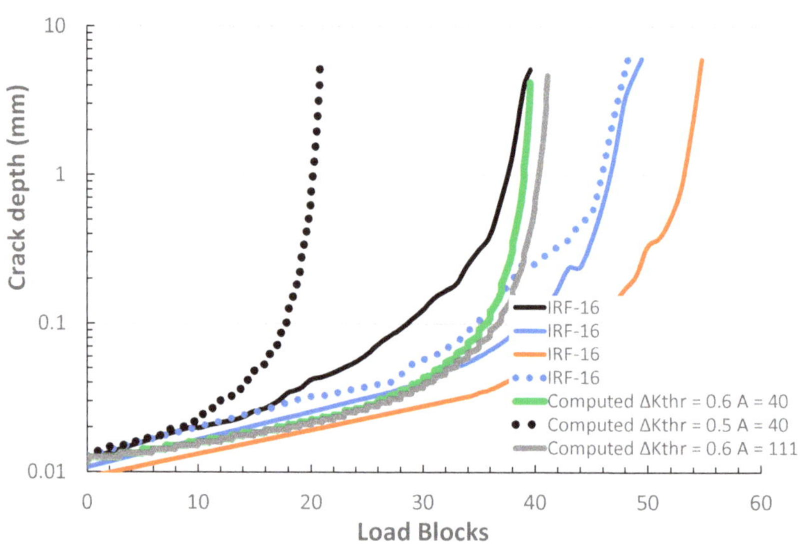

Figure 8. The measured and computed crack depth histories for the helicopter flight load spectrum IRF-E16.

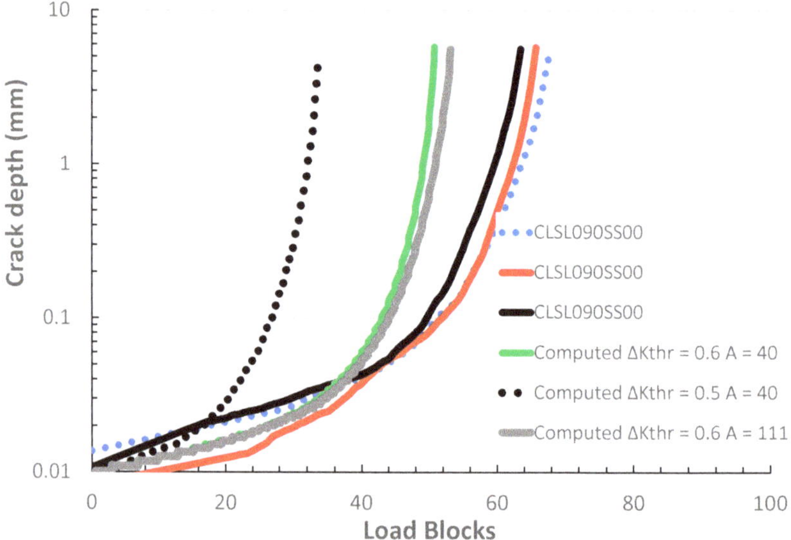

Figure 9. The measured and computed crack depth histories for the helicopter flight load spectrum CLS0900SS00.

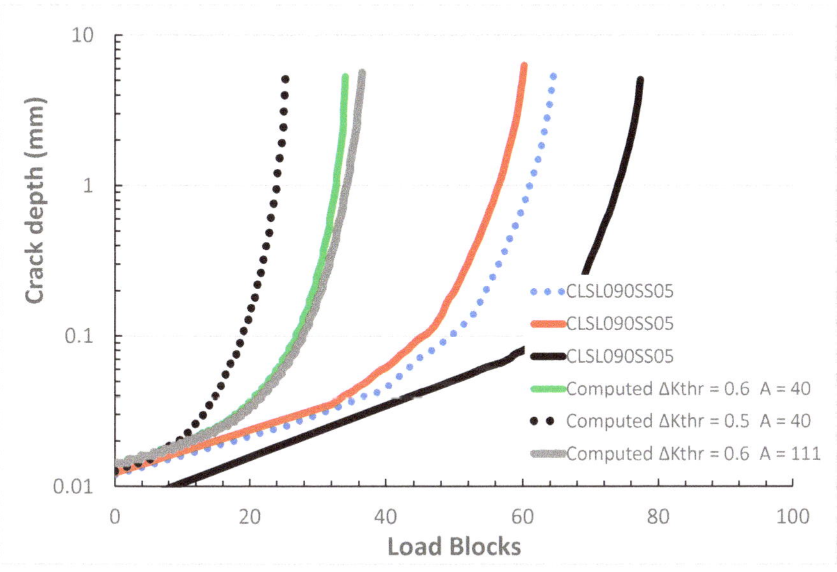

Figure 10. The measured and computed crack depth histories for the helicopter flight load spectrum CLS0900SS05.

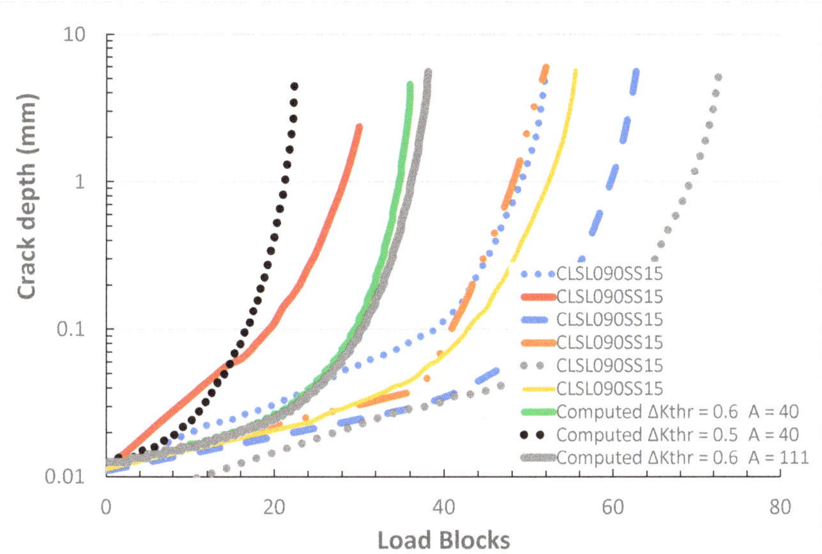

Figure 11. The measured and computed crack depth histories for the helicopter flight load spectrum CLS0900SS15.

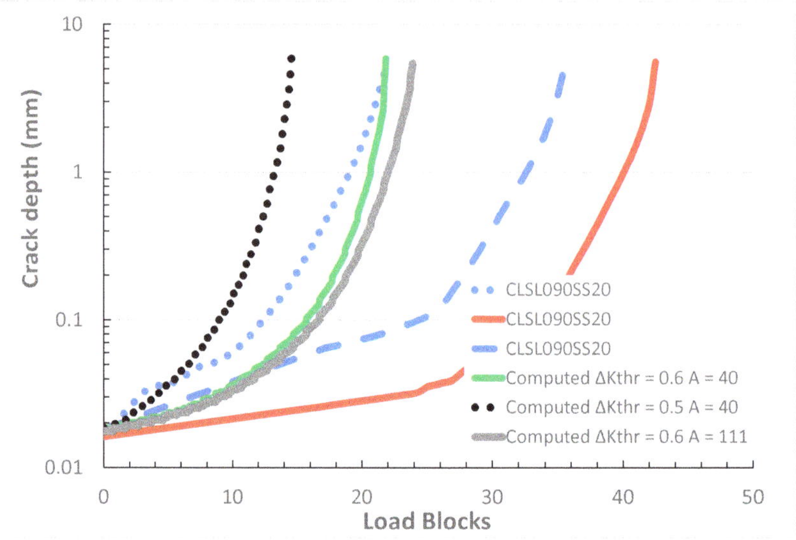

Figure 12. The measured and computed crack depth histories for the helicopter flight load spectrum CLS0900SS20.

5. Material Variability

The variability in crack growth that can arise from a fatigue test was first highlighted by a paper by Virkler et al. [47] This study presented the results of more than sixty constant amplitude R = 0.2 tests on identical 2024-T3 panels which had a constant initial half crack length of 9 mm (see Figure 13). Figure 2 illustrates the variability in crack growth seen in tests on long cracks tested under an operational flight load spectra. Unfortunately, for small cracks the variability in the crack depth histories can be significantly greater than that seen in the long crack curves shown in Figures 2 and 13 [16,48,49]. (The effect of (local) material variability on the growth of small cracks is compounded by the fact that the size and shape of the initial material discontinuity is variable, and generally cannot be tightly controlled [49–51].) The variability in the crack depth history associated with spectra CSL090SS15 is a good example of this, see Figure 11 that presents the variability in the crack depth histories associated with six different cracks.

This raises the question: How much greater would the variability in the crack growth histories shown in Figures 6–12 have been if significantly more tests been performed?

Whilst it is not possible to definitively answer this question, it may be possible to shed some light on the problem space by investigating the effect of small changes in the fatigue threshold on the computed crack growth histories. Given that more than 90% of the Black Hawk flight load spectrum consists of small amplitude loads [45], and that the variability in the growth of small cracks can often be captured by allowing for variability in the term ΔK_{thr} (see Section 3.1 and [2,15,16,18,38,43]) it is anticipated that a small change in ΔK_{thr} should result in a much greater change in the crack growth history. To investigate this hypothesis we reanalysed the various test spectra using $\Delta K_{thr} = 0.5$ MPa $\sqrt{m}$ and $A = 111$ MPa $\sqrt{m}$. The resultant crack depth histories are also shown in Figures 6–12 where they are labelled "Computed $\Delta K_{thr} = 0.5$ $A = 111$". Here we see that, as expected, when values of $\Delta K_{thr} = 0.5$ MPa $\sqrt{m}$ and $A = 111$ MPa $\sqrt{m}$ are used there is a significant reduction in the computed fatigue lives, when compared to the lives computed using $\Delta K_{thr} = 0.6$ MPa $\sqrt{m}$ and $A = 111$ MPa $\sqrt{m}$, and that the computed fatigue lives are now conservative. Bearing in mind that for small cracks growing under combat, maritime, and civil aircraft flight load spectra, it has been shown that the variability in the crack growth histories is captured by allowing for (relatively small) changes in ΔK_{thr}—this appears to suggest that in order to evaluate the effect of simplifying a spectrum, so as to reduce test time, on the fastest possible (lead) crack a statistically significant number of tests should be performed. This requirement is highlighted in Section 3.2.19.1 of the US Joint Services Structural Guidelines [9] that states: "The allowable structural properties shall include all applicable statistical variability".

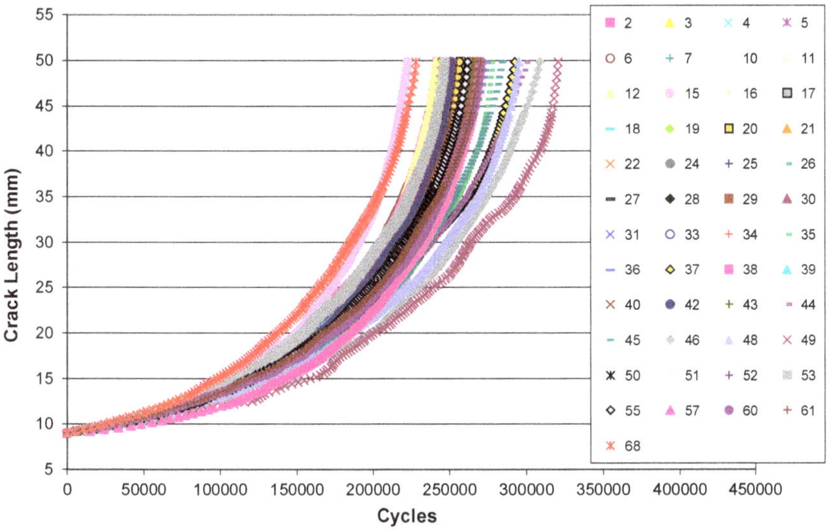

Figure 13. The variability in the crack length histories reported in [46].

To further investigate the variability (scatter) seen in the ASSIST tests let us examine the data associated with test spectra IRF-15 and CSL090SS15, which had information on the largest number of cracks (six). The mean lives, standard deviation (σ), and the projected worst case (mean-3σ) lives are given in Table 3. Here we see that the standard deviation in the test lives is a significant proportion of the mean life. It should be noted that whilst the mean-3σ life and the standard deviation calculations are based on small sample statistics, they nevertheless indicate the need for data on the growth of a greater number of cracks, i.e., additional testing.

Table 3. The values of A and ΔK_{thr} and A used in Figure 14.

Measured/Computed	IRF-E15 Load Blocks to Failure	CSL090SS15 Load Blocks to Failure
Smallest test life	38.3	32.0
Longest test life	66.1	72.7
Mean life	46.8	54.5
Standard deviation (σ)	10.4	13.6
Mean-3σ	15.6	13.8
Computed $\Delta K_{thr} - 0.3$ MPa $\sqrt{m}$, $A = 40$ MPa $\sqrt{m}$	8.9	12.7
Computed $\Delta K_{thr} = 0.3$ MPa $\sqrt{m}$, $A = 111$ MPa $\sqrt{m}$	9.7	13.2

It has been suggested [18] that for small cracks in aluminium alloys the threshold term lies in the range $0.1 < \Delta K_{thr} < 0.3$. Consequently, for spectra IRF-15 and CSL090SS15 the analysis was repeated using $\Delta K_{thr} = 0.3$ MPa $\sqrt{m}$, and either $A = 40$ MPa $\sqrt{m}$, or $A = 111$ MPa $\sqrt{m}$. Comparisons between the measured and computed crack depth histories for these spectra are shown in Figures 14 and 15. The (computed) number of cycles to failure obtained using $A = 40$ MPa $\sqrt{m}$ and $A = 111$ MPa $\sqrt{m}$ are given in Table 3. Here we see that Table 3 and Figures 14 and 15 reveal that the computed crack depth history is a weak function of the cyclic fracture toughness. We also see that when using $\Delta K_{thr} = 0.3$ MPa $\sqrt{m}$ the resultant computed crack depth histories have a near exponential shape. Furthermore, the computed lives to failure are more conservative than the "Mean-3σ" lives as determined from the "limited" number of tests. It is thus suggested that in the absence of a statistically significant number of small crack tests the crack depth curve calculated using the value $\Delta K_{thr} = 0.3$ MPa $\sqrt{m}$ may be a reasonable first estimate for this "worst case" curve.

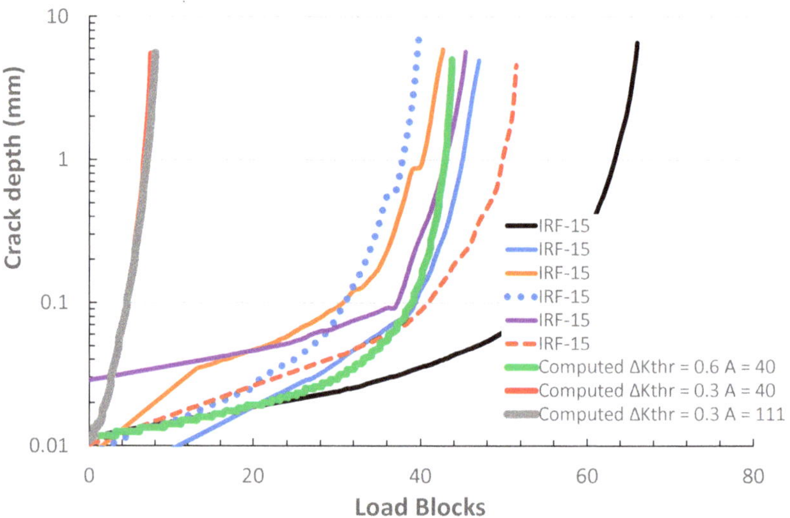

Figure 14. The effect of different toughness on the measured and computed curves for the helicopter flight load spectra IRF-E15.

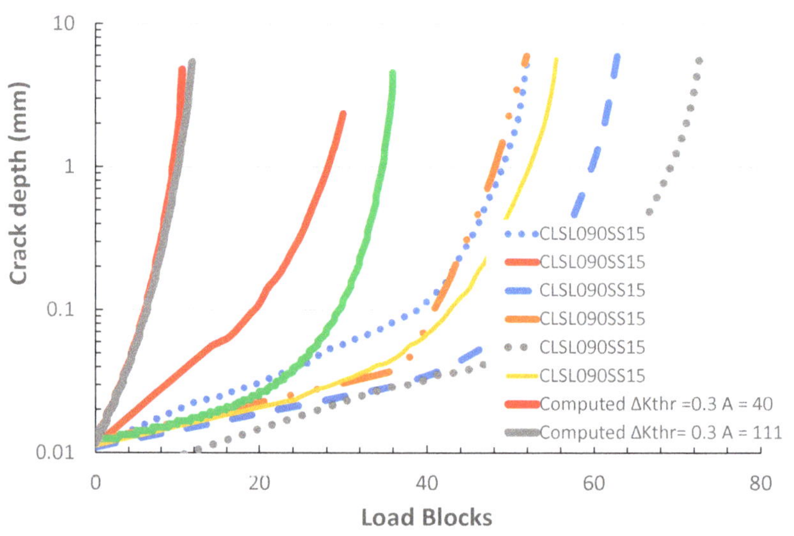

Figure 15. The effect of different toughness on the measured and computed curves for the helicopter flight load spectra CSL090SS15.

On the Shape of the Crack Depth History Curves

In the previous section, we remarked that using $\Delta K_{thr} = 0.3$ MPa $\sqrt{m}$ gave crack depth histories that had a near exponential shape. It was also suggested that, in the absence of a statistically significant number of small crack tests, the crack depth curve computed using $\Delta K_{thr} = 0.3$ MPa $\sqrt{m}$ may be a reasonable first estimate for the worst-case crack depth history. In this context, it should be noted that the USAF Durability Design Handbook [52] explains that the growth of small "lead" cracks, i.e., the fastest growing cracks in an airframe or a component [53,54], in military aircraft is generally exponential. Indeed, this exponential crack growth model is contained within the USAF approach to assessing the risk of failure [55]. This feature, i.e., the exponential growth of lead cracks growing under flight load spectra, was independently validated in [49,56] and is discussed in more detail in [2,49]. However, examining Figures 6–12, we see that the crack growth histories are not exponential. This observation raises an additional question, viz:

If significantly more tests had been performed would the "worst-case" crack depth versus load blocks curves have been (approximately) exponential?

6. Conclusions

The assessment of the economic lives of operational metallic helicopter airframes requires a durability analysis in which the EIDS are sub mm, typically 0.254 mm. Unfortunately, whereas several studies into the ability of crack growth models to perform a damage tolerance analysis of helicopter components subjected to a representative flight load spectrum have been performed few, if any, studies can be found on the ability of crack growth models to perform a valid durability assessment of a component subjected to an operational helicopter flight load spectrum. In this context, the present study has found that the HS equation is able to reasonably accurately compute the growth of small naturally occurring cracks in AA7075-T7351 under several simplified/reduced Black Hawk flight load spectra. This suggests that the HS equation may have the potential to address the question of how to simplify measured spectra in order to reduce the time and complexity of full-scale helicopter fatigue tests.

It is also suggested that, given the inherent variability seen in small crack growth, any round robin test on small cracks, and any test program performed to the effect of spectrum truncation on the growth of small cracks should involve a statistically significant number of tests.

Author Contributions: Conceptualization and methodology—R.J.; analysis of the FALSTAFF loading test case—P.H.; development and validation of the computer software and the helicopter test analysis—D.P.; review, editing, and supervision of D.P. at Monash—R.K.S.R. All authors have read and agreed to the published version of the manuscript.

Funding: This research received no external funding.

Acknowledgments: The authors wish to acknowledge Madeline Burchill at DST for the development of the ASSIST Round Robin Helicopter Program, and Beau Krieg at DST for providing the associated crack growth data. The authors would also like to acknowledge the assistance of Weiping HU at DST who provided the data on the growth of cracks in 7075-T7351 under a FALSTAFF spectrum.

Conflicts of Interest: The authors declare no conflict of interest.

References

1. Lincoln, J.W.; Melliere, R.A. Economic life determination for a military aircraft. *J. Aircr.* **1999**, *36*, 737–742. [CrossRef]
2. Jones, R. Fatigue crack growth and damage tolerance. *Fatigue Fract. Eng. Mater. Struct.* **2014**, *37*, 463–483. [CrossRef]
3. MIL-STD-1530D, Department of Defense Standard Practice Aircraft Structural Integrity Program (ASIP). 13 October 2016. Available online: http://everyspec.com/MIL-STD/MIL-STD.../download.php?spec=MIL-STD-1530D (accessed on 2 February 2020).
4. ASTM E647-13a. *Measurement of Fatigue Crack Growth Rates*; ASTM: West Conshohocken, PA, USA, 2013.

5. Newman, J.C.; Irving, P.; Lin, J.; Le, D. Crack growth in a complex helicopter component under spectrum loading. *Fatigue Fract. Eng. Mater. Struct.* **2006**, *29*, 949–958. [CrossRef]

6. Irving, P.; Lin, J.; Bristow, J. Damage tolerance in helicopters—Report on the round-robin challenge. In Proceedings of the American Helicopter Society 59th Annual Forum, Phoenix, AR, USA, 6–8 May 2003; Available online: https://vtol.org/store/product/a-round-robin-exercise-to-assess-capability-topredict-crack-growth-lives-and-inspection-intervals-for-damage-tolerant-design-in-helicopters-4262.cfm (accessed on 29 May 2020).

7. Peng, D.; Jones, R.; Sinha, A.; Mathews, N.; Singh Raman, R.K.; Phan, N.T.; Nguyen, T. Analysis of fatigue crack growth in a helicopter component. In Proceedings of the Asian/Australian Rotorcraft Forum 2018, Seogwipo City, Jeju Island, Korea, 30 October–1 November 2018; Available online: https://vtol.org/arf2018 (accessed on 29 May 2020).

8. Cansdale, R.; Perrett, B. The Helicopter Damage Tolerance Round Robin Challenge. In Proceedings of the Workshop on Fatigue Damage of Helicopters, Pisa, Italy, 12–13 September 2002.

9. Department of Defense Joint Service Specification Guide, Aircraft Structures, JSSG-2006. October 1998. Available online: http://everyspec.com/USAF/USAF-General/JSSG-2006_10206/ (accessed on 10 July 2020).

10. Structures Bulletin EZ-SB-19-01, Durability and Damage Tolerance Certification for Additive Manufacturing of Aircraft Structural Metallic Parts, Wright Patterson Air Force Base, OH, USA. 10 June 2019. Available online: https://daytonaero.com/usaf-structures-bulletins-library/ (accessed on 2 February 2020).

11. Burchill, M. Advancing Structural Simulation to Drive Innovative Sustainment Technologies: ASSIST Helicopter Crack Growth Prediction Challenge Definition. 18 June 2019. Available online: https://www.researchgate.net/publication/341790926_Approved_for_Public_Release_Advancing_Structural_Simulation_to_drive_Innovative_Sustainment_Technologies_ASSIST_Helicopter_Crack_Growth_Prediction_Challenge_Definition?channel=doi&linkId=5ed4f110458515294527add2&showFulltext=true (accessed on 16 May 2020).

12. ASSIST Challenge 3. Available online: https://www.researchgate.net/publication/342201234_Spectra (accessed on 16 May 2020).

13. Hartman, A.; Schijve, J. The effects of environment and load frequency on the crack propagation law for macro fatigue crack growth in aluminium alloys. *Eng. Fract. Mech.* **1970**, *1/4*, 615–631. [CrossRef]

14. Lo, M.; Jones, R.; Bowler, A.; Dorman, M.; Edwards, D. Crack growth at fastener holes containing intergranular cracking. *Fatigue Fract. Eng. Mater. Struct.* **2017**, *40*, 1664–1675. [CrossRef]

15. Tamboli, D.; Barter, S.; Jones, R. On the growth of cracks from etch pits and the scatter associated with them under a miniTWIST spectrum. *Int. J. Fatigue* **2018**, *109*, 10–16. [CrossRef]

16. Jones, R.; Molent, L.; Barter, S. Calculating crack growth from small discontinuities in 7050-T7451 under combat aircraft spectra. *Int. J. Fatigue* **2013**, *55*, 178–182. [CrossRef]

17. Main, B.; Evans, R.; Walker, K.; Yu, X.; Molent, L. Lessons from a Fatigue Prediction Challenge for an Aircraft Wing Shear Tie Post. *Int. J. Fatigue* **2019**, *123*, 53–65. [CrossRef]

18. Jones, R.; Singh Raman, R.K.; McMillan, A.J. Crack growth: Does microstructure play a role? *Eng. Fracture Mech.* **2018**, *187*, 190–210. [CrossRef]

19. Zhang, Y.; Zheng, K.; Heng, J.; Zhu, J. Corrosion-Fatigue Evaluation of Uncoated Weathering Steel Bridges. *Appl. Sci.* **2019**, *9*, 3461. [CrossRef]

20. Godefroid, L.B.; Moreira, L.P.; Vilela, T.C.G.; Faria, G.L.; Candido, L.C.; Pinto, E.S. Effect of chemical composition and microstructure on the fatigue crack growth resistance of pearlitic steels for railroad application. *Int. J. Fatigue* **2019**, *120*, 241–253. [CrossRef]

21. Cano, A.J.; Salazar, A.; Rodríguez, J. Evaluation of different crack driving forces for describing the fatigue crack growth behaviour of PET-G. *Int. J. Fatigue* **2018**, *107*, 27–32. [CrossRef]

22. Jones, R.; Peng, D.; Michopoulos, J.G.; Kinloch, A.J. Requirements and Variability Affecting the Durability of Bonded Joints. *Materials* **2020**, *23*, 1468. [CrossRef]

23. Jones, R.; Hu, W.; Kinloch, A.J. A convenient way to represent fatigue crack growth in structural adhesives. *Fatigue Fract. Eng. Mater. Struct.* **2015**, *38*, 379–391. [CrossRef]

24. Jones, R.; Kinloch, A.J.; Hu, W. Cyclic-fatigue crack growth in composite and adhesively-bonded structures: The FAA slow crack growth approach to certification and the problem of similitude. *Int. J. Fatigue* **2016**, *88*, 10–18. [CrossRef]

25. Rocha, A.V.M.; Akhavan-Safar, A.; Carbas, R.; Marques, E.A.S.; Goyal, R.; El-Zein, M.; da Silva, L.F.M. Paris law relations for an epoxy-based adhesive. *Proc. Inst. Mech. Eng. Part L J. Mater. Des. Appl.* **2019**. [CrossRef]
26. Brunner, A.J.; Stelzer, S.; Pinter, G.; Terrasi, G.P. Cyclic fatigue delamination of carbon fiber-reinforced polymer-matrix composites: Data analysis and design considerations. *Int. J. Fatigue* **2016**, *83*, 293–299. [CrossRef]
27. Chocron, T.; Banks-Sills, L. Nearly mode I fracture toughness and fatigue delamination propagation in a multidirectional laminate fabricated by a wet-layup. *Phys. Mesomech.* **2019**, *22*, 107–140. [CrossRef]
28. Simon, I.; Banks-Sills, L.; Fourman, V. Mode I delamination propagation and R-ratio effects in woven composite DCB specimens for a multi-directional layup. *Int. J. Fatigue* **2017**, *96*, 237–251. [CrossRef]
29. Jones, R.; Kinloch, A.J.; Michopoulos, J.G.; Brunner, A.J.; Phan, N. Delamination growth in polymer-matrix fibre composites and the use of fracture mechanics data for material characterisation and life prediction. *Compos. Struct.* **2017**, *180*, 316–333. [CrossRef]
30. Yao, L.; Alderliesten, R.C.; Jones, R.; Kinloch, A.J. Delamination fatigue growth in polymer-matrix fibre composites: A methodology for determining the design and lifing allowables. *Compos. Struct.* **2018**, *196*, 8–20. [CrossRef]
31. Jones, R.; Stelzer, S.; Brunner, A.J. Mode I, II and Mixed Mode I/II delamination growth in composites. *Compos. Struct.* **2014**, *110*, 317–324. [CrossRef]
32. Simon, I.; Banks-Sills, L.; Fourman, V.; Eliasi, R. Delamination Propagation and Load Ratio Effects in DCB MD Woven Composite Specimens. *Procedia Struct. Integr.* **2016**, *2*, 205–212. [CrossRef]
33. Schwalbe, K.H. On the beauty of analytical models for fatigue crack propagation and fracture—A personal historical review. In *Fatigue and Fracture Mechanics: 37th Volume*; ASTM International: West Conshohocken, PA, USA, 2011; pp. 3–73. [CrossRef]
34. Jones, R.; Michopoulos, J.G.; Iliopoulos, A.P.; Raman, R.S.; Phan, N.; Nguyen, T. Representing crack growth in additively manufactured Ti-6Al-4V. *Int. J. Fatigue* **2018**, *116*, 610–622. [CrossRef]
35. Jones, R.; Sing Raman, R.K.; Iliopoulos, A.P.; Michopoulos, J.; Phan, N.; Peng, D. Additively manufactured Ti-6Al-4V replacement parts for military aircraft. *Int. J. Fatigue* **2019**, *124*, 227–235. [CrossRef]
36. Iliopoulos, A.P.; Jones, R.; Michopoulos, J.; Phan, N.; Singh Raman, R.K. Crack growth in a range of additively manufactured aerospace structural materials. *Aerospace* **2018**, *5*, 118. [CrossRef]
37. Iliopoulos, A.P.; Jones, R.; Michopoulos, J.G.; Phan, N.; Rans, C. Further Studies into Crack Growth in Additively Manufactured Materials. *Materials* **2020**, *13*, 2223. [CrossRef]
38. Jones, R.; Molaei, R.; Fatemi, A.; Peng, D.; Phan, N. A note on computing the growth of small cracks in AM Ti-6Al-4V. In Proceedings of the 1st Virtual European Conference on Fracture (VECF1), 29 June 2020; Available online: https://www.youtube.com/watch?v=W8rTAREK7ak&feature=youtu.be (accessed on 31 May 2020).
39. Kundu, S.; Jones, R.; Peng, D.; Matthews, N.; Alankar, A.; Singh Raman, R.K.; Huang, P. Review of requirements for the durability and damage tolerance certification of additively manufactured aircraft structural parts and AM repairs. *Materials* **2020**, *13*, 1341. [CrossRef]
40. Under Secretary, Acquisition and Sustainment, Directive-Type Memorandum (DTM)-19-006-Interim Policy and Guidance for the Use of Additive Manufacturing (AM) in Support of Materiel Sustainment, Pentagon, Washington, DC. 21 March 2019. Available online: https://www.esd.whs.mil/Portals/54/Documents/DD/issuances/dtm/DTM-19-006.pdf?ver=2019-03-21-075332-443 (accessed on 2 February 2020).
41. Zhuang, W.; Liu, Q.; Hu, W. The effect of specimen thickness on fatigue crack growth rate and threshold behaviour in aluminium alloy 7075-T7351. In Proceedings of the 6th Australasian Congress on Applied Mechanics (ACAM 6), Perth, Australia, 12–15 December 2010.
42. Jones, R.; Lo, M.; Peng, D.; Bowler, A.; Dorman, M.; Janardhana, M.; Iyyer, N.S. A study into the interaction of intergranular cracking and cracking at a fastener hole. *Meccanica* **2015**, *50*, 517–532. [CrossRef]
43. Molent, L.; Jones, R. The influence of cyclic stress intensity threshold on fatigue life scatter. *Int. J. Fatigue* **2016**, *82*, 748–756. [CrossRef]
44. Liao, M.; Renaud, G.; Bombardier, Y. Short/Small Crack Model Development for Aircraft Structural Life Assessment. In Proceedings of the 13th International Conference on Fracture, Beijing, China, 16–21 June 2013.
45. Krake, L. Helicopter Airframe Fatigue Spectra Generation. *Adv. Mater. Res.* **2014**, *891–892*, 720–725. [CrossRef]

46. Peng, D.; Huang, P.; Jones, R. Practical computational fracture mechanics for aircraft structural integrity. In *Aircraft Sustainment and Repair*; Jones, R., Baker, A., Matthews, N., Champagne, V., Eds.; Elsevier: Oxford, UK, 2017; ISBN 9780081005408.

47. Virkler, D.A.; Hillberry, B.M.; Goel, P.K. The statistical nature of fatigue crack propagation. *Trans. ASME* **1979**, *101*, 148–153. [CrossRef]

48. Huynh, J.; Molent, L.; Barter, S. Experimentally derived crack growth models for different stress concentration factors. *Int. J. Fatigue* **2008**, *30*, 1766–1786. [CrossRef]

49. Wanhill, R.H.; Barter, S.; Molent, L. *Fatigue Crack Growth Failure and Lifing Analyses for Metallic Aircraft Structures and Components*; Springer: Dordrecht, The Netherlands, 2019; ISBN 978-94-024-1673-2.

50. Molent, L.; Dixon, B. Airframe metal fatigue revisited. *Int. J. Fatigue* **2020**, 131. [CrossRef]

51. Barter, S.A.; Molent, L.; Wanhill, R.H. Typical fatigue-initiating discontinuities in metallic aircraft structures. *Int. J. Fatigue* **2012**, *41*, 1–198. [CrossRef]

52. Manning, S.D.; Yang, Y.N. USAF Durability Design Handbook: Guidelines for the Analysis and Design of Durable Aircraft Structures, AFWAL-TR-83-3027, Flight Dynamics Directorate, Wright Laboratory, Air Force Systems Command, Wright-Patterson Air Force Base. January 1984. Available online: https://apps.dtic.mil/dtic/tr/fulltext/u2/a206286.pdf (accessed on 16 May 2020).

53. Hover, P.W.; Berens, A.P.; Loomis, J. Update of the Probability of Fracture (PROF) Computer Program for Aging Aircraft Risk Analysis, AFRL-VA-WP-TR-1999-3030, Flight Dynamics Directorate, Wright Laboratory, Air Force Systems Command, Wright-Patterson Air Force Base. November 1998. Available online: https://apps.dtic.mil/dtic/tr/fulltext/u2/a363010.pdf (accessed on 16 May 2020).

54. Molent, L.; Barter, S.A.; Wanhill, R.J.H. The lead crack fatigue lifing framework. *Int. J. Fatigue* **2011**, *33*, 323–331. [CrossRef]

55. Berens, A.P.; Hovey, P.W.; Skinn, D.A. Risk Analysis for Aging Aircraft Fleets—Volume 1: Analysis, WL-TR-91-3066, Flight Dynamics Directorate, Wright Laboratory, Air Force Systems Command, Wright-Patterson Air Force Base. October 1991. Available online: https://apps.dtic.mil/dtic/tr/fulltext/u2/a252000.pdf (accessed on 16 May 2020).

56. Molent, L.; Barter, S.A. A comparison of crack growth behaviour in several full-scale airframe fatigue tests. *Int. J. Fatigue* **2007**, *9*, 1090–1099. [CrossRef]

Article

Fatigue and Crack Growth under Constant- and Variable-Amplitude Loading in 9310 Steel Using "Rainflow-on-the-Fly" Methodology

James C. Newman, Jr.

Department of Aerospace Engineering, Mississippi State University, Mississippi, MS 39762, USA; newmanjr@ae.msstate.edu; Tel.: +1-(901)-734-6642

Abstract: Fatigue of materials, like alloys, is basically fatigue-crack growth in small cracks nucleating and growing from micro-structural features, such as inclusions and voids, or at micro-machining marks, and large cracks growing to failure. Thus, the traditional fatigue-crack nucleation stage (N_i) is basically the growth in microcracks (initial flaw sizes of 1 to 30 μm growing to about 250 μm) in metal alloys. Fatigue and crack-growth tests were conducted on a 9310 steel under laboratory air and room temperature conditions. Large-crack-growth-rate data were obtained from compact, C(T), specimens over a wide range in rates from threshold to fracture for load ratios (R) of 0.1 to 0.95. New test procedures based on compression pre-cracking were used in the near-threshold regime because the current ASTM test method (load shedding) has been shown to cause load-history effects with elevated thresholds and slower rates than steady-state behavior under constant-amplitude loading. High load-ratio (R) data were used to approximate small-crack-growth-rate behavior. A crack-closure model, FASTRAN, was used to develop the baseline crack-growth-rate curve. Fatigue tests were conducted on single-edge-notch-bend, SEN(B), specimens under both constant-amplitude and a Cold-Turbistan+ spectrum loading. Under spectrum loading, the model used a "Rainflow-on-the-Fly" subroutine to account for crack-growth damage. Test results were compared to fatigue-life calculations made under constant-amplitude loading to establish the initial microstructural flaw size and predictions made under spectrum loading from the FASTRAN code using the same micro-structural, semi-circular, surface-flaw size (6-μm). Thus, the model is a unified fatigue approach, from crack nucleation (small-crack growth) and large-crack growth to failure using fracture mechanics principles. The model was validated for both fatigue and crack-growth predictions. In general, predictions agreed well with the test data.

Keywords: fatigue; crack growth; metallic materials; plasticity; crack closure; spectrum loading

Citation: Newman, J.C., Jr. Fatigue and Crack Growth under Constant- and Variable-Amplitude Loading in 9310 Steel Using "Rainflow-on-the-Fly" Methodology. *Metals* **2021**, *11*, 807. https://doi.org/10.3390/met11050807

Academic Editor: Filippo Berto

Received: 17 April 2021
Accepted: 13 May 2021
Published: 15 May 2021

Publisher's Note: MDPI stays neutral with regard to jurisdictional claims in published maps and institutional affiliations.

1. Introduction

This article is dedicated to Dr. Wolf Elber and his remarkable achievements.

On 12 January 2019, Wolf took off on his final flight. His passion was to soar in his glider over the Blue Ridge Mountains in southwest Virginia, USA. With his passing,

the fatigue and fracture mechanics community lost a great pioneer. Dr. Elber had a distinguished career at the NASA Langley Research Center, and later as head of the U.S. Army Research Laboratory at Langley. His professional accolades are immortalized by his discovery of the "plasticity induced crack closure" phenomenon [1,2]. Thank you, Wolf, for your friendship and brilliant mind over the years. We will greatly miss you.

Fatigue of metallic materials is divided into several phases: crack nucleation, small- and large-crack growth, and fracture [3]. Crack nucleation is controlled by local stress and strain concentrations and is associated with cyclic slip-band formation from dislocation movement leading to intrusions and extrusions [4,5]. Although cyclic slip may be necessary in pure metals, the presence of inclusions, voids, or machining marks in metal alloys greatly affects the crack-nucleation process. (Herein, small-crack growth includes microstructurally and physically small cracks.) The small-crack growth regime is the growth in cracks from inclusions, voids, or machining marks, ranging from 1 to 30 μm in depth [6]. Schijve [7] has shown that, for polished surfaces of pure metals and commercial alloys, the formation of a small crack of about 100 μm in size can consume 60 to 80% of the fatigue life. This is the reason that there is so much interest in the growth behavior of small cracks. Large-crack growth and failure are regions where fracture-mechanics parameters have been successful in correlating and predicting fatigue-crack growth and fracture. In the past three decades, fracture-mechanics concepts have also been successful in predicting the growth in small cracks under constant-amplitude and spectrum-loading using crack-closure theory [6].

The engineered metallic materials are inhomogeneous and anisotropic when viewed on a microscopic scale. For example, these materials are composed of an aggregate of small grains, inclusion particles or voids. These inclusion particles are of different chemical compositions to the bulk material, such as silicate or alumina inclusions in steels. Because of their nonuniform microstructure, local stresses may be concentrated at these locations and may cause the initiation of fatigue cracks. Crack initiation is primarily a surface phenomenon because: (1) local stresses are usually highest at the surface, (2) an inclusion particle of the same size has a higher stress concentration at the surface than in the interior, (3) the surface is subjected to adverse environmental conditions, and (4) the surfaces are susceptible to inadvertent damage. The growth in "naturally" initiated cracks in commercial aluminum alloys has been investigated by Bowles and Schijve [8], Morris et al. [9] and Kung and Fine [10]. In some cases, small cracks initiated at inclusions and the Stage I period of crack growth were eliminated [6]. This tendency toward inclusion initiation rather than slip-band (Stage I) cracking was found to depend on stress level and inclusion content [10]. Similarly, defects (such as tool marks, scratches and burrs) from manufacturing and service-induced events will also promote initiation and Stage II crack growth [6].

During the last three decades, test and analysis programs on "small-crack" behavior have shown that majority of the fatigue life is consumed by small-crack growth from a micro-structural feature for a variety of metal alloys [11–14]. The smallest measured flaw sizes using plastic-replica methods ranged from 10 to 30 μm for aluminum alloys (2024-T3; 7075-T6), aluminum–lithium alloy (2090-T8ED41) and 4340 steel; and the crack-propagation life was about 90 percent of the fatigue life for constant-amplitude and spectrum loading. Thus, a large portion of nucleation life (N_i) in classical fatigue is small-crack growth, from a micro-structural feature to about 250 μm for these materials. The exception was the Ti-6Al-4V alloy [13], where the smallest measured flaw sizes were about 20 μm and crack-growth lives were about 50% of the fatigue life. However, small-crack analyses of similar Ti-6Al-4V specimens machined from two engine discs [15,16] showed that the initial flaw sizes from 2 to 20 μm predicted the scatter band in fatigue life on open-hole fatigue tests quite well. Therefore, the crack-growth approach provides a unified theory for the determination of fatigue lives for these materials. However, for pure- and single-crystal materials, nucleation cycles are required to transport dislocations at critical locations, develop slip bands, and cracks.

Fatigue-crack growth under variable-amplitude and spectrum loading is composed of complex crack-shielding mechanisms (plasticity, roughness and fretting debris) and damage accumulation due to cyclic plastic deformations around a crack front in metallic materials. Typically, "rainflow" methods [17] are applied to variable-amplitude loading to develop a load sequence that is used to compute damage accumulation and life because damage relations are, generally, a non-linear function of the crack-driving parameters. However, crack-tip damage is only a function of the current loading and load history (material memory). Loading in the future has no bearing on the "current" damage. The life-prediction code, FASTRAN [18], has a "rainflow-on-the-fly" methodology [19] to compute damage as the cyclic load history is applied, and there was no need to reorder the spectra. Herein, several variable-amplitude loading sequences were developed to test and to validate the "rainflow-on-the-fly" subroutine.

The paper presents the results of large-crack-growth-rate tests conducted on compact, C(T), specimens made of 9310 steel [20] over a wide range of constant-amplitude loading to establish the baseline crack-growth-rate curve for fatigue and crack-growth analyses. Compression pre-cracking methods [21–26] were used to generate test data in the near-threshold regime because the ASTM E-647 test method [27,28] using the load-shedding procedure was shown to cause load-history effects and slower crack-growth rates than steady-state behavior [25,26]. Both compression pre-cracking constant-amplitude (CPCA) and load-reduction (CPLR) methods were used. A crack-closure analysis was used to collapse the rate data from C(T) specimens into a narrow band over many orders of magnitude in rates using a plane-strain constraint factor for low rates and modeled a constraint-loss regime to plane-stress behavior at high rates. For steels, small- and large-crack data (without load-history effects) tend to agree well [29]. Thus, the high-R large-crack data in the near-threshold regime is a good estimate for small-crack behavior, as proposed by Herman et al. [30]. A Two-Parameter Fracture Criterion [31] was used to characterize the fracture behavior. Fatigue tests were conducted on 9310 steel single-edge-notch-bend, SEN(B), specimens [32] under both constant-amplitude and a modified Cold-Turbistan [33] spectrum loading. The test results were compared to the life calculations or predictions made using the FASTRAN code.

2. Material and Specimen Configurations

The Boeing Company (Seattle, WA, USA) provided a 9310 steel rod (150 mm diameter by 950 mm length) to Mississippi State University. C(T), SEN(B), and tensile specimens ($B = 6.35$ mm) were machined in the longitudinal direction (cracks perpendicular to longitudinal direction) and heat-treated by special procedures [20]. The yield stress, σ_{ys}, was 980 MPa, ultimate tensile strength, σ_u, was 1250 MPa, and modulus of elasticity was $E = 208.6$ GPa. Figure 1 shows C(T) and SEN(B) specimens with back-face strain (BFS) gauges used to monitor crack growth in the C(T) specimens and crack nucleation and growth in the SEN(B) specimen.

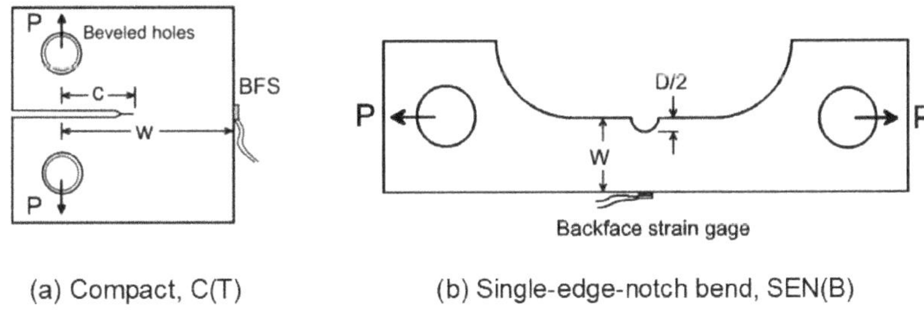

(a) Compact, C(T)　　　　　(b) Single-edge-notch bend, SEN(B)

Figure 1. Fatigue-crack-growth and fatigue specimens tested and analyzed.

3. "Rainflow-on-the-Fly" Methodology

Since the Paris–Elber non-linear power law is used to compute crack-growth damage in the FASTRAN life-prediction code [18], a "rainflow-on-the-fly" subroutine was developed to compute crack growth under variable-amplitude loading. Damage calculations are illustrated in Figure 2 and they are based on calculations of stress amplitudes above the crack-opening stress, S_o. Damage only occurs during the loading amplitude but unloading may change crack-tip deformations and affect the subsequent damage during the next loading amplitude. Crack-opening stress, S_o, is calculated at the minimum stress, S_1, and Δc_1 is a function of $\Delta S_{eff} = S_2 - S_o$. The ΔS_{eff} value is used to compute ΔK_{eff} and then the crack-growth rate (dc/dN) per loading amplitude, which gives Δc_1. The next unloading to S_3 did not close the crack, and the next maximum loading was to S_4. Here, damage is $\Delta c_2 = f(S_4 - S_o) - \Delta c_1 + f(S_2 - S_3)$, which captures the larger damage due to S_4. The total damage is $\Delta c = \Delta c_1 + \Delta c_2$. Again, each stress range is used to calculate an effective stress-intensity factor and then the corresponding crack extension from the crack-growth-rate relation. Application of minimum stress, S_5, caused the crack surfaces to close, and rainflow-on-the-fly logic was reset. Total damage, Δc, captures the essence of rainflow logic and applies damage in the proper sequence using a cycle-by-cycle calculation. (Note that a cycle is defined as any minimum–maximum–minimum stress.)

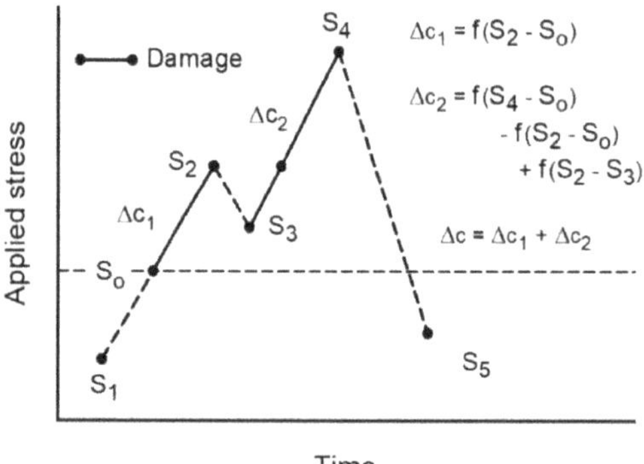

Figure 2. Damage calculations using crack-closure theory under variable-amplitude loading.

Herein, several specially designed spectra were developed to exercise the FASTRAN code and calculations are shown for these spectra. Each spectrum had sequences of stress ranges that would require "rainflow" logic to calculate the proper crack extension. Comparisons are made between: (1) linear damage, (2) rainflow logic and (3) FASTRAN. Linear damage is using only the stress range to calculate crack extension and ignoring load interactions. Rainflow logic is a manual calculation of crack extension using the basic rainflow concept principles to compute the proper crack extension. FASTRAN uses the "rainflow-on-the-fly" subroutine to remember stress history and to compute the proper damage as the crack-opening stress in FASTRAN was held constant, as specified, and the crack-growth relation was a simple Paris–Elber relation as

$$dc/dN = 5 \times 10^{-10}(\Delta K_{eff})^3 \tag{1}$$

where $\Delta K_{eff} = \Delta S_{eff}\sqrt{\pi c}$. The spectra were applied to an infinite plate under remote uniform stress, S, with an initial crack half-length (c_i) of 5 mm.

Spectrum A is a Christmas-Tree type loading sequence and is shown in Figure 3a and was designed to severely test the Rainflow-on-the-Fly option in FASTRAN. Without Rainflow analyses, the difference in crack-growth lives for the sequence would be several orders-of-magnitude in error using linear damage [19]. One block of loading is defined from point A to B, and the sequence was repeated until the final crack length was reached or the specimen failed. The analytical crack-closure model in FASTRAN was turned off by intentionally setting the crack-opening stress, S_o, to a constant value of 50 MPa.

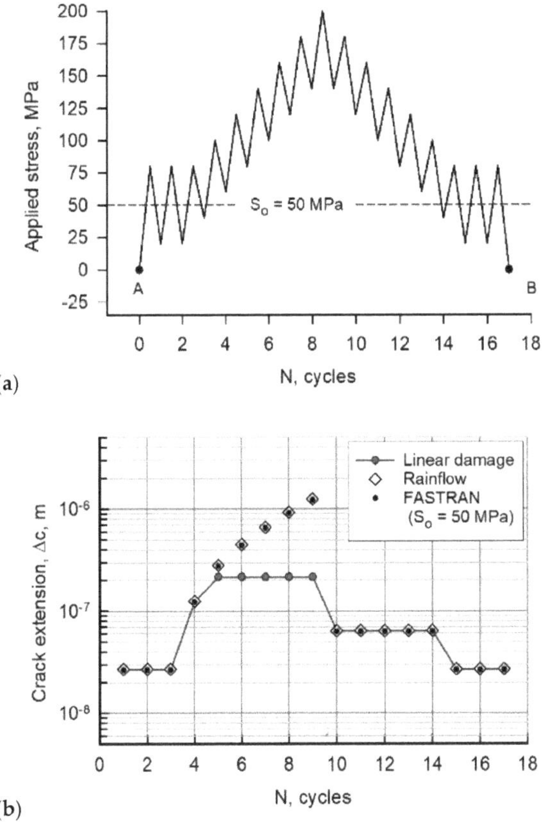

Figure 3. (**a**) Loading sequence A. (**b**) Calculated crack extension per applied cycle for load sequence A.

Figure 3b shows crack-growth increments calculated for each loading amplitude. If the crack-opening stress is greater than the minimum stress, S_{min}, then $\Delta S_{eff} = S_{max} - S_o$. If S_{min} is greater than the crack-opening stress, then $\Delta S_{eff} = S_{max} - S_{min}$. These equations apply only for "linear damage" (lines with a solid blue symbol) and for loadings that do not require Rainflow methods. Diamond symbols show what a Rainflow method would give in crack extension. (Note that a cycle is defined as any minimum to maximum to minimum loading, and crack-growth damage only occurs during loading. The unloading part of the cycle causes reverse plastic deformations that may affect damage during the next load application.) FASTRAN with the Rainflow subroutine gives the solid (black) symbols that exactly match what the Rainflow method predicted.

The second spectrum (B) was designed after some of the European standard spectra with various stress amplitudes that have a number of constant-amplitude cycles, and the spectrum had a number of cycles that go from maximum–minimum–maximum or minimum–maximum–minimum loading. Some cycles had a Christmas-Tree type loading that was interrupted with other stress amplitudes. Again, this spectrum does require rainflow logic to calculate the correct damage.

The first case (B1) had a constant crack-opening stress (50 MPa) for the complete spectrum, as shown in Figure 4a. The calculated crack extension using linear damage, rainflow and FASTRAN are shown in Figure 4b. Out of the 57 cycles, 18 cycles required rainflow logic to compute the correct crack extension. FASTRAN agreed with the Rainflow predicted crack extensions.

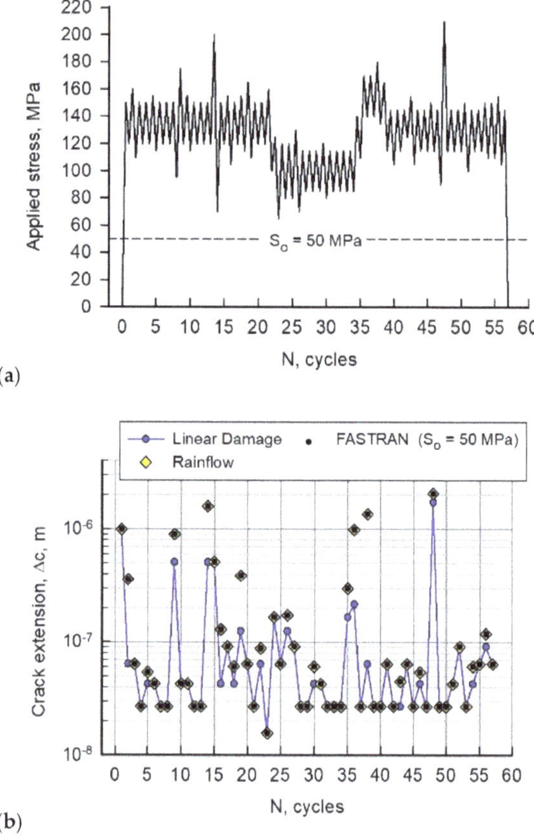

Figure 4. (**a**) Loading sequence B1. (**b**) Calculated crack extension per applied cycle for load sequence B1.

In the FASTRAN code, crack-opening stresses will change as a function of stress history. Thus, to simulate a changing crack-opening stress, the second case (B2) had a constant crack-opening stress of 50 MPa for the first 22 cycles and then S_o = 90 MPa for the remainder of loading, as shown in Figure 5a. Of course, the first 22 cycles have the same behavior as Case B1, but now cycles 23 to 34 do not require rainflow logic ($S_{min} < S_o$) and crack extensions were computed directly from the Paris–Elber relation (Equation (1)). These crack-extension comparisons are shown in Figure 5b. However, the crack extensions during cycles 35 to 37 were much less than in Case B1 due to the higher crack-opening stress and lower ΔK_{eff} values.

Figure 6 shows the calculated crack-opening stresses from FASTRAN (α = 2) using the full crack-closure model for load sequence B using the cycle-by-cycle option. The results for Block 1 are shown as the dashed (red) lines. The crack-opening stresses are calculated at a minimum applied stress and remain constant until the crack closes again at the next minimum applied stress. In many cycles, the crack-tip region was open ($S_{min} > S_o$) during the applied stress amplitudes. The model starts with a "zero" crack-opening stress and the opening stresses increase as the crack grows and leaves plastically deformed material in the wake of the crack tip.

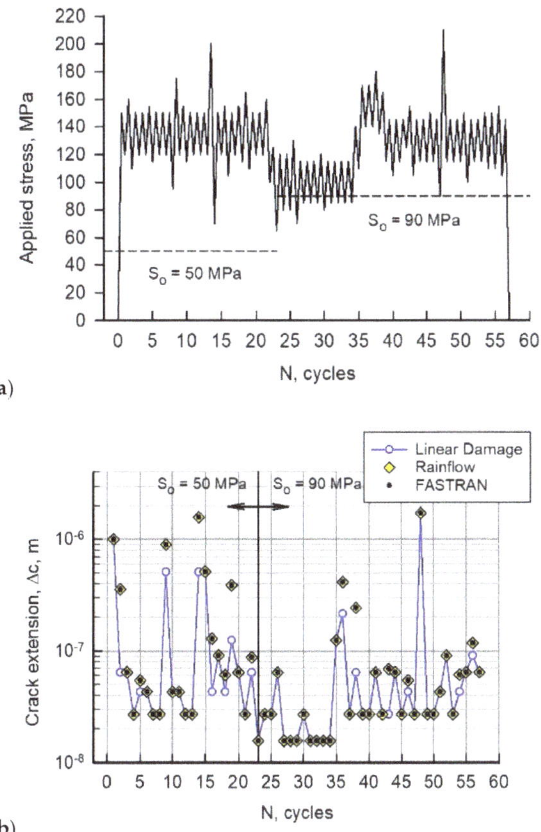

(a)

(b)

Figure 5. (**a**) Loading sequence B. (**b**) Calculated crack extension per applied cycle for load sequence B2.

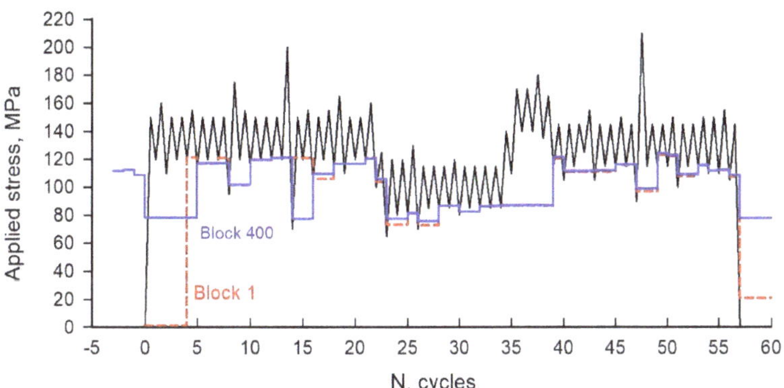

Figure 6. Loading sequence B and calculated crack-opening stresses.

Calculated crack-opening-stress history for Block 400, at about 110,000 cycles (half-life), are shown as solid (blue) lines. Here, the cycle numbers are the same cycles as in Block 1 (first sequence). The calculated crack-opening stresses where in agreement over most of the sequence, except the only major difference was in cycles 1 to 5, where the crack surfaces were fully open. However, the crack extension during the first five cycles in Block 1 was a factor of 8 larger than that during the 400th block because of the higher crack-opening stress. Most of the damage from cycles 1 to 5 was from the loading on cycle 1.

4. Fatigue-Crack-Growth and Fracture Tests on 9310 Steel

Compact, C(T), specimens were used to generate the ΔK against rate (dc/dN) data on the 9310 steel at room temperature and 20 Hertz [20] over a wide range in stress ratios (R = 0.1 to 0.95). These ΔK-rate data are shown in Figure 7. Tests were conducted from near threshold to fracture. A BFS gauge was used to monitor crack growth and to measure crack-opening loads using the compliance-offset method. In the low-rate regime, compression pre-cracking constant amplitude (CPCA) and load reduction (CPLR) methods were used to generate the ΔK-rate data. In general, the test frequency was dropped to about 5 to 8 Hertz as the cracks were grown to failure. The data show the normal spread with the R value but shows more spread in the threshold and fracture regions. In the fracture region, the spread is due to the fracture being controlled by the maximum stress-intensity factor and would not collapse on a ΔK-rate plot. For example, the stress-intensity-factor range at failure, $\Delta K_c = K_{Ie}$ *(1 − R)*, where K_{Ie} is the elastic fracture toughness. Thus, high R tests would fracture at a lower ΔK_c value than low R tests. The spread in the threshold region is suspected to be due to the load-history effects caused by the load-reduction procedure used in the CPLR method. Load-reduction tests, as specified in ASTM E-647 [28], have been shown [25,26] to induce more spread in ΔK in the threshold region than the mid-rate region. The spread has been associated with a rise in the crack-closure behavior during load-reduction tests [34,35].

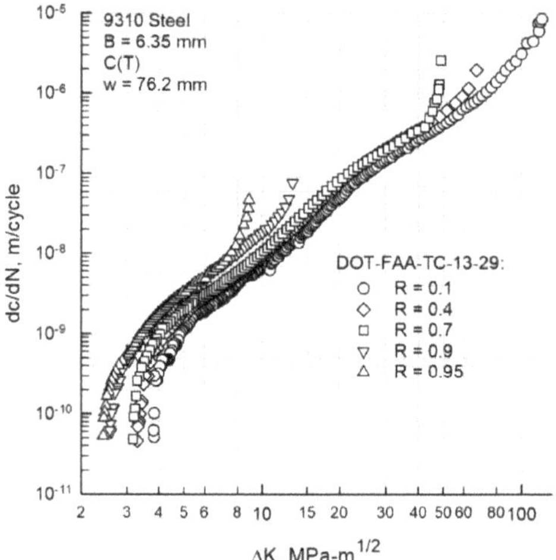

Figure 7. Stress-intensity-factor range against rate for 9310 steel over range in stress ratio (R).

To evaluate the fracture toughness of the material, only the fatigue-crack-growth tests were available. In most tests, the cracks in the C(T) specimens were grown to failure under cyclic loading. Here, the final recorded crack length and the maximum fatigue load were used to calculate K_{Ie} (elastic stress-intensity factor at failure), and these results are shown in Figure 8, as solid (blue) circular symbols. Most tests failed at large c_i/w ratios (>0.7). For comparison, similar test results conducted on a 4340 steel [36] are shown as open circular symbols. Here, one 4340 steel specimen was fatigue cracked to a lower crack-length-to-width ratio and statically pulled to failure (solid square symbol).

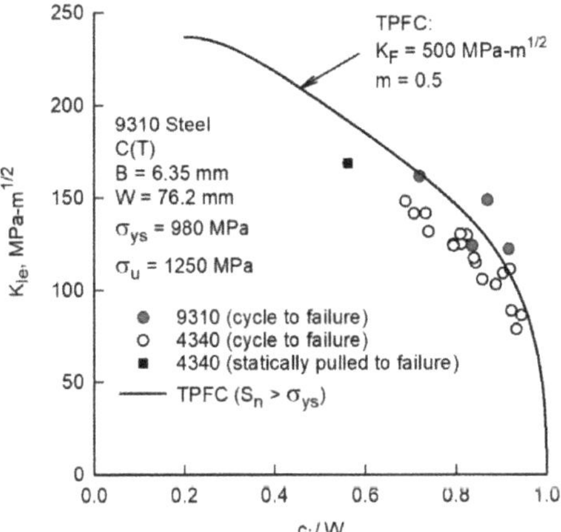

Figure 8. Elastic stress-intensity factor at failure on 9310 and 4340 steel C(T) specimens.

The solid curve in Figure 8 was based on the Two-Parameter Fracture Criterion (TPFC). A fracture criterion was derived [37] that accounted for the elastic-plastic behavior of a cracked material based on notch-strength analysis. The criterion was based on expressing the elastic stress-intensity factor at failure in terms of net-section stress as

$$K_{Ie} = P_f / \sqrt{BW}\, F = S_n \sqrt{\pi c}\, F_n \tag{2}$$

where F_n is the usual boundary-correction factor based on net-section stress. Equations for S_n and F_n are given in [37] for the C(T) specimen. The fracture criterion is

$$K_F = K_{Ie} / \Phi \tag{3}$$

$$\Phi = 1 - m\, S_n / S_u \tag{4}$$

$$\Phi = \left(\sigma_{ys} / S_n \right) \left(1 - m\, S_n / S_u \right) \tag{5}$$

where K_F and m are the two material fracture parameters. The stress S_u (ultimate value of elastic net-section stress) was computed from the load required to produce a fully plastic region or hinge on the net section based on the ultimate tensile strength, σ_u. For the compact specimen, S_u is a function of load eccentricity and is $1.62\sigma_u$ for c/w = 0.5 [37]. The fracture parameters, K_F and m, are assumed to be constant in the same sense as the ultimate tensile strength; that is, the parameters may vary with material thickness, state of stress, temperature, and rate of loading. If m is equal to zero in Equations (3) and (4), then K_F is equal to the elastic stress-intensity factor at failure, and the equation represents the behavior of low-toughness (brittle) materials under plane-strain behavior. If m is equal to unity, the equation represents fracture behavior of high-toughness materials (plane-stress fracture).

For given fracture-toughness parameters, K_F, and m, the elastic stress-intensity factor at failure is

$$K_{Ie} = K_F / \left(1 + 2m\gamma / \sigma_{ys} \right) \tag{6}$$

$$K_{Ie} = \left\{ \sqrt{(m\gamma)^2 + 2\gamma S_u} - m\gamma \right\} \sqrt{\pi c}\, F_n \tag{7}$$

$$K_{Ie} = S_u \sqrt{\pi c}\, F_n \tag{8}$$

where $\gamma = K_F \sigma_{ys} / \left[2\, S_u \sqrt{\pi c}\, F_n \right]$. In 1973, a relationship between K_F/E and m [37] was found for many materials and crack configurations. Herein, the relationship was used to help select an m value. As the 9310 steel exhibited a high toughness, an m-value of 0.5 was selected. The corresponding K_F value was 500 MPa-$\sqrt{m}$ to fit the 9310 steel fracture data. Solid curve in Figure 8 was calculated from the TPFC for the 76.2 mm wide compact specimens. These calculations show how the initial crack length affects the elastic fracture toughness. In addition, the variation in specimen width (not shown) would greatly affect the elastic fracture toughness, in that larger width specimens produce large K_{Ie} values.

5. Fatigue-Crack-Growth and Crack-Closure Analyses

Crack-opening-stress equations for constant-amplitude loading were developed from an early analytical crack-closure model (S_o) calculations [38]. As the number of elements within the plastic-zone region in the model was increased to 20 [18] and the crack-growth increment was modelled on a cycle-by-cyclic basis, new S_o equations were made for a single crack in a very wide plate under uniform remote applied stress, S. The new set of equations was developed to fit the results from the revised closure model and, again, gave S_o as a function of stress ratio (R), maximum stress level (S_{max}/σ_0) and the constraint factor (α). The new equations are

$$S_o / S_{max} = A_0 + A_1 R + A_2 R^2 + A_3 R^3$$

$$(9)$$

$$S_o / S_{max} = A_0 + A_1 R$$

$$(10)$$

where $R = S_{min}/S_{max}$, $S_{max} < 0.8\, \sigma_o$, and $S_{min} > -\sigma_o$. The A_i coefficients are functions of α and S_{max}/σ_o and are given by

$$A_0 = \left(0.9453 - 0.514\, \alpha + 0.1355\, \alpha^2 - 0.0133\, \alpha^3 \right) \left[\cos(\beta) \right]^{(0.8\, \alpha - 0.1)} \qquad (11)$$

$$\beta = \pi S_{max} / (2\alpha \sigma_o)$$

$$A_1 \left(= 0.5719 - 0.1726\, \alpha + 0.019\, \alpha^2 \right) S_{max} / \sigma_o \qquad (12)$$

$$A_2 = 0.975 - A_0 - A_1 - A_3 \qquad (13)$$

$$A_3 = 2A_0 + A_1 - 1 \qquad (14)$$

A crack-closure analysis was then performed on the fatigue-crack growth (ΔK-rate) data from C(T) specimens in Figure 7 to determine the ΔK_{eff}-rate relation. The K-analogy concept [18] was used to calculate the crack-opening stresses (or loads) for C(T) specimens from the above equations. The ΔK_{eff}-rate data are shown in Figure 9. Selection of the lower constraint factor, 2.5, was found to reasonably collapse the ΔK-rate data into an almost unique relation.

In the threshold region, the lower R tests exhibited a rise in crack-opening loads as the ΔK level was reduced in a load-reduction test. Even the CPLR method showed a load-history effect, but not as much as the current ASTM procedure [28]. The upper constraint factor, 1.15, and constraint-loss range was selected to help fit spectrum crack-growth tests [20]. The lower vertical dashed line at $(\Delta K_{eff})_{th}$ is the estimated threshold for the steel [39], and the upper vertical dashed line at $(\Delta K_{eff})_T$ is the location of constraint loss from plane-strain to plane-stress behavior [40]. The solid (blue) lines with circular (yellow) symbols shows the baseline crack-growth-rate curve for FASTRAN. Fatigue-crack-growth, fracture and tensile properties for the 6.35-mm thick 9310 steel are given in Table 1.

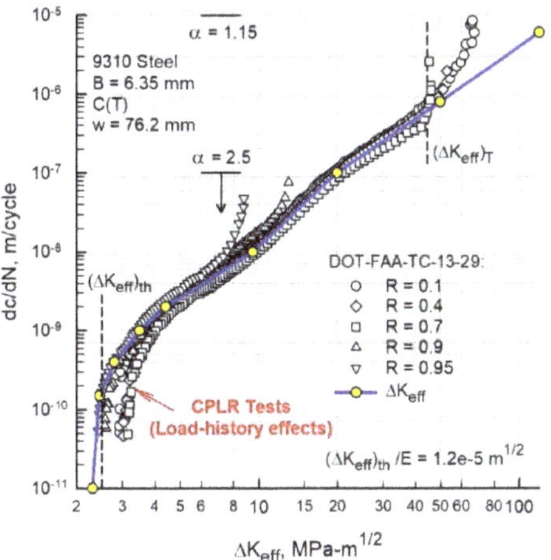

Figure 9. Effective stress-intensity factor range against rate for 9310 steel C(T) specimens.

Table 1. Effective stress-intensity factor range against rate, fracture and tensile properties for 9310 steel (B = 6.35 mm).

ΔK_{eff}, (MPa-m$^{1/2}$)	da/dN and dc/dN (m/cycle)	Crack-Growth, Fracture and Tensile Properties
2.30	1.0×10^{-11}	$\Delta K_o = 0\ q = 6$
2.45	1.5×10^{-10}	$\alpha_1 = 2.5$ at 1.0×10^{-7} m/cycle
2.80	4.0×10^{-10}	$\alpha_2 = 1.15$ at 1.0×10^{-5} m/cycle
3.50	1.0×10^{-9}	$K_F = 500$ MPa-m$^{1/2}$
4.40	2.0×10^{-9}	$m = 0.5$
9.50	1.0×10^{-8}	$\sigma_{ys} = 980$ MPa
20.0	1.0×10^{-7}	$\sigma_u = 1250$ MPa
50.0	8.0×10^{-7}	$\sigma_o = 1115$ MPa
120.0	6.0×10^{-6}	$E = 208.6$ GPa

6. Fatigue Behavior of Notched Specimens

For most fatigue-crack-growth analyses, linear-elastic analyses have been found to be adequate. The linear-elastic effective stress-intensity factor range developed by Elber [2] is given by

$$\Delta K_{eff} = (S_{max} - S_o)\ \sqrt{\pi c}\ F(c/w) \tag{15}$$

where S_{max} is maximum stress, S_o is crack-opening stress, and $F(c/w)$ is the boundary correction factor. However, for high-stress intensity factors and low-cycle fatigue conditions, linear-elastic analyses are inadequate and nonlinear crack-growth parameters are needed. To account for plasticity, a portion of the Dugdale [41] cyclic-plastic-zone size (ω) has been added to the crack length, c. The cyclic-plastic-zone-corrected effective stress-intensity factor [42,43] is

$$(\Delta K_p)_{eff} = (S_{max} - S_o)\ \sqrt{\pi d}\ F(d/w) \tag{16}$$

where $d = c + \omega/4$ and $F(d/w)$ is the cyclic-plastic-zone-corrected boundary-correction factor. The cyclic plastic zone is given by

$$\omega = (1 - R_{eff})^2\ \rho/4 \tag{17}$$

where $R_{eff} = S_o/S_{max}$ and plastic-zone size (ρ) for a crack in a large plate [41] is

$$\rho = c\ \{\sec[\pi S_{max}/(2\alpha\sigma_o)] - 1\} \tag{18}$$

where flow stress, σ_o, is multiplied by the constraint factor (α). Herein, the cyclic-plastic-zone corrected effective stress-intensity factor range (Equation (17)) will be used in the fatigue-life predictions.

The FASTRAN life-prediction code [18] was used to model crack growth from an initial micro-structural flaw size to failure and the crack-growth relation used is

$$dc/dN = C_{1i}[(\Delta K_p)_{eff}]^{C_{2i}} \left\{ 1 - [\Delta K_o/(\Delta K_p)_{eff}]^p \right\} / [1 - (K_{max}/K_{Ie})^q] \tag{19}$$

where C_{1i} and C_{2i} are coefficient and exponent for each linear segment (i = 1 to n), respectively. The $(\Delta K_p)_{eff}$ is cyclic-plastic-zone corrected effective stress-intensity factor, ΔK_o is effective threshold, K_{max} is maximum stress-intensity factor, K_{Ie} is elastic fracture toughness (which is, generally, a function of crack length, specimen width, and specimen type), p and q are constants selected to fit test data in either the threshold or fracture regimes, respectively. Herein, no threshold was modeled and ΔK_o was set equal to zero; thus, p was not needed. Near-threshold behavior was modeled with the multi-linear equation (independent of R). Fracture was modeled using the TPFC (K_F and m) [31,37].

Fatigue tests were conducted on SEN(B) specimens under: (1) constant-amplitude loading (R = 0.1) and (2) Cold-Turbistan+ loading. The semi-circular edge notch was chemically polished. The Cold-Turbistan+ spectrum was obtained from [33], adding a constant load to make the overall R = 0.1 (compression loads not allowed on SEN(B) specimens). Figure 10 shows the constant-amplitude tests plotting σ_{max} (notch root elastic stress) against cycles to failure (N_f). The square symbols are single-edge-notch tension, SEN(T), specimens [44], whereas the solid circular symbols are from the current study. A trial-and-error procedure was used to find the 6-μm semi-circular surface flaw at the center of the notch to fit the fatigue data. Herein, the 6-μm flaw is considered an equivalent initial flaw size (EIFS) to fit the S-N behavior under constant-amplitude loading. The solid curve shows calculated lives that used available crack-growth-rate data (rates $\geq 4 \times 10^{-11}$ m/cycle) for $\sigma_{max} \geq 980$ MPa. The dashed (blue) curve used the estimated ΔK_{eff}-rate relation for rates below 4×10^{-11} m/cycle (see Figure 9). On Figure 10, the horizontal dashed (black) line is where $\Delta K_{eff} = (\Delta K_{eff})_{th} = 2.3$ MPa$\sqrt{m}$. Upper and lower bound calculations were made with 4- and 10-μm, respectively.

Figure 11 shows the test data (solid circular symbols) under the Cold-Turbistan+ spectrum. The open symbols are the retest of the two runout tests, but at higher applied notch elastic stress. FASTRAN predictions using the same 6-μm surface flaw fell at the lower bound of the test data using the baseline curve from Figure 9 (Table 1). If a threshold of 2.3 MPa$\sqrt{m}$ (no crack growth) was selected, then the cycles to failure were near the upper bound of the test data. Selecting a ΔK_{eff}-rate relation between the vertical dashed line and the baseline (blue) curve below 10^{-10} m/cycle, would have given a more accurate predicted fatigue strength. These results indicate that the selection of the baseline curve in Figure 9 for rates below 10^{-10} m/cycle is very important. In addition, in the fatigue endurance-limit region, there could also be some build-up of fretting oxide debris on the small-crack surfaces, which could increase the fatigue life by elevating the crack-opening loads. Further study is required to include fretting debris on the crack surfaces in the FASTRAN model.

To study the crack-closure behavior and life predictions under the Cold-Turbistan+ spectrum, the predicted crack-opening-stress ratios are shown in Figure 12 for a small portion of the spectrum (first 80 cycles). Here, the FASTRAN code used the full crack-closure model with a constraint factor of 2.5. Calculated crack-opening stresses began at the minimum stress in the spectrum and the crack-opening stress increases as the crack grows and leaves plastically deformed material in the wake of the crack tip. Calculations were also made at Block 35 (about half-life) and show that the crack-opening stresses are, generally, higher than Block 1. These results are somewhat surprising, in that the predicted crack-opening stresses were generally at the minimum applied stress values for the larger cyclic amplitudes. This would imply that the loading cycle would be fully effective in growing the crack. Thus, the Cold-Turbistan+ spectrum would be classified as an "accelerating" spectra.

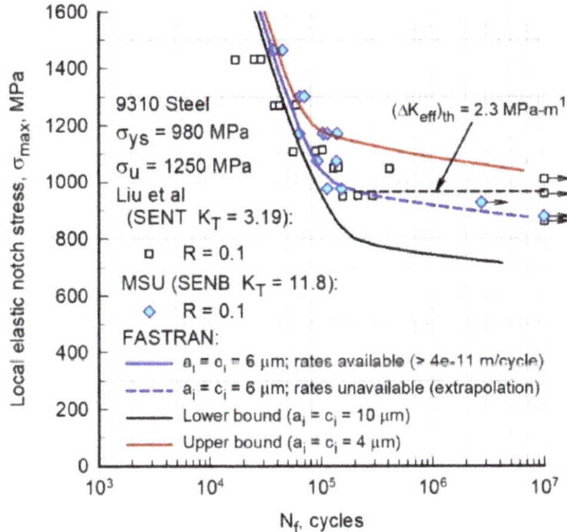

Figure 10. Measured and calculated stress-life behavior under constant-amplitude loading.

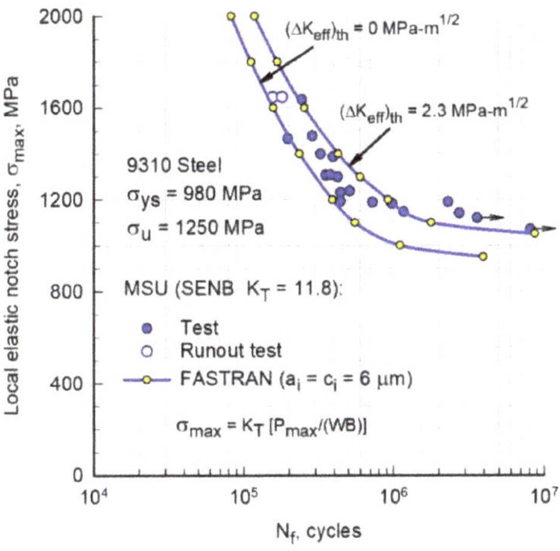

Figure 11. Measured and predicted stress-life behavior under Cold-Turbistan+ spectrum loading.

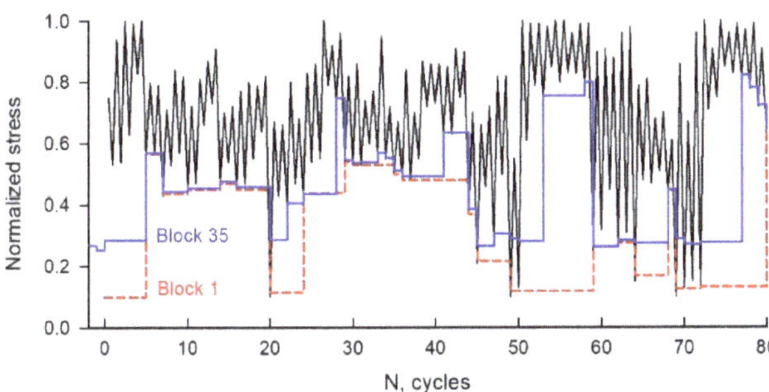

Figure 12. Part of Cold Turbistan+ spectrum loading and calculated crack-opening-stress ratios.

These results are important, in that Rainflow analyses for fatigue-crack growth are a function of load history. Fortunately, the current Rainflow analyses, used in the literature, would be conservative without considering crack-closure load-history effects. However, the accuracy of fatigue-crack-growth calculations under spectrum loading would be greatly improved with "Rainflow-on-the-Fly" methodology.

7. Concluding Remarks

Fatigue of materials, like alloys, is basically the fatigue-crack growth in small cracks nucleating at micro-structural features, such as inclusions and voids, or at micro-machining marks, and large cracks growing to failure. Thus, the traditional fatigue-crack nucleation stage (N_i) is basically the growth in microcracks (initial flaw sizes of 1 to 30 μm growing to about 250 μm) in a variety of metal alloys. Large-crack growth and failure are regions where fracture-mechanic parameters were successful in correlating and predicting fatigue-crack growth and fracture. In the last three decades, fracture-mechanics concepts have also been successful in predicting "fatigue" (growth of small cracks) under constant-amplitude and spectrum loading using crack-closure theory. Therefore, the crack-growth approach provides a unified theory for the determination of fatigue lives for metal alloys. However, for pure- and single-crystal materials, there are nucleation cycles required to transport dislocations at critical locations, develop slip bands, and cracks.

Tests were conducted on compact, C(T), and single-edge-notch-bend, SEN(B), specimens made of a 9310 steel (B = 6.35 mm) under laboratory air and room temperature conditions. The C(T) crack-growth specimens were tested over a wide range in stress ratios (R = 0.1 to 0.95) and crack-growth rates from threshold to fracture. A crack-closure model (FASTRAN) was used to collapse the ΔK-rate data onto an almost unique ΔK_{eff}-rate relation over more than four orders-of-magnitude in rates.

Fatigue tests were conducted on the SEN(B) specimens under constant-amplitude loading (R = 0.1) and a Cold-Turbistan+ spectrum loading. The ΔK_{eff}-rate crack-growth relation was used to calculate or predict fatigue behavior on the SEN(B) specimens using the crack-closure model and small-crack theory. The constant-amplitude fatigue tests were used to determine an initial semi-circular surface flaw size (6-μm) to fit the test data. The 6-μm flaw is considered an equivalent initial flaw size (EIFS). Using the same initial flaw size enabled the FASTRAN code to predict the fatigue behavior under the Cold-Turbistan+ spectrum loading quite well. Rainflow-on-the-Fly methodology was validated on a complex spectrum loading and indicated that the calculated damage was a function of load history and that the usual Rainflow methods would not capture correct crack-growth damage, unless the method was updated during crack-growth history.

Funding: This research received no external funding.

Conflicts of Interest: The author declares no conflict of interest.

Nomenclature

a	crack depth in thickness direction, mm
a_i	initial crack depth in thickness direction, mm
B	specimen thickness, mm
c	crack length in width direction, mm
c_i	initial crack length in width direction, mm
D	single-edge-notch diameter, mm
da/dN	crack-growth rate in depth direction, m/cycle
dc/dN	crack-growth rate in width direction, m/cycle
E	modulus of elasticity, GPa
F	boundary-correction factor
F_n	boundary-correction factor based on net-section stress

K	stress-intensity factor, MPa$\sqrt{\text{m}}$
K_F	elastic-plastic fracture toughness, MPa$\sqrt{\text{m}}$
K_{Ie}	elastic fracture toughness, MPa$\sqrt{\text{m}}$
K_{max}	maximum stress-intensity factor, MPa$\sqrt{\text{m}}$
K_T	elastic stress-concentration factor
m	fracture toughness parameter
N	number of cycles
N_f	number of cycles to failure
P	applied load, kN
P_f	failure load, kN
P_{max}	maximum applied load, kN
P_{min}	minimum applied load, kN
R	load ratio (P_{min}/P_{max})
S	applied remote stress, MPa
S_{max}	maximum applied stress, MPa
S_{min}	minimum applied stress, MPa
S_o	crack-opening stress, MPa
w	specimen width, mm
α	tensile constraint factor
ΔK	stress-intensity factor range, MPa$\sqrt{\text{m}}$
ΔK_c	stress-intensity-factor range at failure, MPa$\sqrt{\text{m}}$
ΔK_{eff}	effective stress-intensity factor range, MPa$\sqrt{\text{m}}$
$(\Delta K_{eff})_{th}$	effective stress-intensity factor range threshold, MPa$\sqrt{\text{m}}$
$(\Delta K_{eff})_T$	effective stress-intensity factor range transition, MPa$\sqrt{\text{m}}$
ΔK_o	effective threshold as a function of R, MPa$\sqrt{\text{m}}$
ρ	plastic-zone size, mm
σ_o	flow stress (average of σ_{ys} and σ_u), MPa
σ_{ys}	yield stress (0.2 percent offset), MPa
σ_u	ultimate tensile strength, MPa
ω	cyclic plastic-zone size, mm

Abbreviations

BFS	back-face strain
CPCA	compression pre-cracking constant amplitude
CPLR	compression pre-cracking load reduction
C(T)	compact (tension) specimen
SEN(B)	single-edge-notch (bend) specimen
SEN(T)	single-edge-notch (tension) specimen

References

1. Elber, W. Some Effects of Crack Closure on the Mechanism of Fatigue Crack Propagation under Cyclic Tension Loading. Ph.D. Thesis, University of New South Wales, Sydney, Australia, 1968.
2. Elber, W. The significance of fatigue crack closure. In *Damage Tolerance in Aircraft Structures*; ASTM STP 486; American Society of Testing Materials: Philadelphia, PA, USA, 1971; pp. 230–242.
3. Schijve, J. *Significance of Fatigue Cracks in Micro-Range and Macro-Range*; ASTM STP 415; American Society of Testing Materials: Philadelphia, PA, USA, 1967; pp. 415–459.
4. Suresh, S. *Fatigue of Materials*; Cambridge University Press: Cambridge, UK, 1991.
5. Stephens, R.I.; Fatemi, A.; Stephens, R.; Fuchs, H.O. *Metal Fatigue in Engineering*; John Wiley & Sons: New York, NY, USA, 2000.
6. Newman, J.C., Jr. The merging of fatigue and fracture mechanics concepts—A historical perspective. *Prog. Aerosp. Sci.* **1998**, *34*, 345–388. [CrossRef]
7. Schijve, J. Four lectures on fatigue crack growth. *Eng. Fract. Mech.* **1979**, *11*, 167–221. [CrossRef]
8. Bowles, C.Q.; Schijve, J. The role of inclusions in fatigue crack initiation in an aluminum alloy. *Int. J. Fract.* **1973**, *9*, 171–179. [CrossRef]
9. Morris, W.; Buck, O.; Marcus, H. Fatigue crack initiation and early propagation in Al 2219-T851. *Metall. Trans.* **1976**, *7*, 1161–1165. [CrossRef]
10. Kung, C.Y.; Fine, M.E. Fatigue crack initiation and microcrack growth in 2024-T4 and 2124-T4 aluminum alloys. *Metall. Trans.* **1979**, *10*, 603–610. [CrossRef]

11. Newman, J.C., Jr.; Edwards, P.R. *Short-Crack Growth Behaviour in an Aluminum Alloy—An AGARD Cooperative Test Programme*; AGARD Report No. 732; North Atlantic Treaty Organization: Neuilly Sur Seine, France, 1988.
12. Mom, A.J.A.; Raizenne, M.D. (Eds.) *AGARD Engine Disc Cooperative Test Programme*; AGARD Report No. 766; North Atlantic Treaty Organization: Neuilly Sur Seine, France, 1988.
13. Edwards, P.R.; Newman, J.C., Jr. (Eds.) *Short-Crack Growth Behaviour in Various Aircraft Materials*; AGARD Report No. 767; North Atlantic Treaty Organization: Neuilly Sur Seine, France, 1990.
14. Newman, J.C., Jr.; Wu, X.R.; Venneri, S.L.; Li, C.G. *Small-Crack Effects in High-Strength Aluminum Alloys—A NASA/CAE Cooperative Program*; NASA RP-1309; National Aeronautics and Space Administration: Washington, WA, USA, 1994.
15. Newman, J.C., Jr. Application of a closure model to predict crack growth in three engine disc materials. *Int. J. Fract.* **1996**, *80*, 193–218. [CrossRef]
16. Newman, J.C., Jr.; Phillips, E.P.; Everett, R.A., Jr. *Fatigue Analyses Under Constant- and Variable-Amplitude Loading Using Small-Crack Theory*; NASA TM-209329; National Aeronautics and Space Administration: Washington, WA, USA, 1999.
17. Murakami, Y. The rainflow method in fatigue. In Proceedings of the International Symposium Fatigue Damage Measurement and Evaluation under Complex Loadings, Fukuoka, Japan, 25–26 July 1991.
18. Newman, J.C., Jr. *FASTRAN—A Fatigue Crack Growth Life Prediction Code Based on the Crack-Closure Concept, Version 5.4 User Guide*; Fatigue and Fracture Associates LLC: Eupora, MS, USA, 2013.
19. Newman, J.C., Jr. Rainflow-on-the-fly methodology: Fatigue-crack growth under aircraft spectrum loading. *Adv. Mater. Res.* **2014**, *891*, 771–776. [CrossRef]
20. Newman, J.C., Jr.; Yamada, Y.; Ziegler, B.M.; Shaw, J.W. *Small- and Large-Crack Databases for Rotorcraft Materials*; DOT/FAA/TC-13/29; Department of Transportation: Washington, DC, USA, 2014.
21. Topper, T.H.; Au, P.; El Haddad, M.L. The effects of compressive overloads on the threshold stress intensity for short cracks. In Proceedings of the AGARD Conference Proceedings No. 328, Toronto, ON, Canada, 19–24 September 1982; pp. 11.1–11.7.
22. Suresh, S. Crack initiation in cyclic compression and its application. *Eng. Fract. Mech.* **1985**, *21*, 453–463. [CrossRef]
23. Pippan, R. The growth of short cracks under cyclic compression. *Fatigue Fract. Eng. Mater. Struct.* **1987**, *9*, 319–328. [CrossRef]
24. Pippan, R.; Plöchl, L.; Klanner, F.; Stüwe, H.P. The use of fatigue specimens pre-cracked in compression for measuring threshold values and crack growth. *ASTM J. Test. Eval.* **1994**, *22*, 98–106.
25. Newman, J.C., Jr.; Schneider, J.; Daniel, A.; McKnight, D. Compression pre-cracking to generate near threshold fatigue-crack-growth rates in two aluminum alloys. *Int. J. Fatigue* **2005**, *27*, 1432–1440. [CrossRef]
26. Newman, J.C., Jr.; Ruschau, J.J.; Hill, M.R. Improved test method for very low fatigue-crack-growth-rate data. *Fatigue Fract. Eng. Mater. Struct.* **2011**, *34*, 270–279. [CrossRef]
27. Hudak, S., Jr.; Saxena, S.; Bucci, R.; Malcolm, R. *Development of Standard Methods of Testing and Analyzing Fatigue Crack Growth Rate Data*; AFML TR 78-40; Materials Laboratory: WPAFB, OH, USA, 1978.
28. ASTM. *Standard Test Method for Measurement of Fatigue Crack Growth Rates*; E-647; ASTM International: West Conshohocken, PA, USA, 2015.
29. Swain, M.H.; Everett, R.A.; Newman, J.C., Jr.; Phillips, E.P. The growth of short cracks in 4340 steel and aluminum-lithium 2090. In *AGARD Report No. 767*; Edwards, P.R., Newman, J.C., Jr., Eds.; North Atlantic Treaty Organization: Neuilly Sur Seine, France, 1990; pp. 7.1–7.30.
30. Herman, W.; Hertzberg, R.; Jaccard, R. A simplified laboratory approach for the prediction of short crack behavior in engineering structures. *Fatigue Fract. Eng. Mater. Struct.* **1988**, *11*, 303–320. [CrossRef]
31. Newman, J.C., Jr. Fracture analysis of various cracked configurations in sheet and plate materials. In *Properties Related to Fracture Toughness*; ASTM STP 605; American Society of Testing Materials: Philadelphia, PA, USA, 1976; pp. 104–123.
32. Ziegler, B.M. Fatigue and Fracture Testing and Analysis on Four Materials. Ph.D. Thesis, Mississippi State University, Mississippi State, MS, USA, April 2011.
33. ten Have, A.A. *Cold Turbistan—Final Definition of a Standardized Fatigue Test Loading Sequence for Tactical Aircraft Cold Section Engine Disc*; NLR TR 87054 L; Nationaal Lucht-en Ruimtevaartlaboratorium: Delft, The Netherlands, 1987.
34. Yamada, Y.; Newman, J.C., Jr. Crack closure under high load-ratio conditions for Inconel 718 near threshold behavior. *Eng. Fract. Mech.* **2009**, *76*, 209–220. [CrossRef]
35. Yamada, Y.; Newman, J.C., Jr. Crack closure behavior of 2324-T39 aluminum alloy near threshold conditions for high load ratio and constant Kmax tests. *Int. J. Fatigue* **2009**, *31*, 1780–1787. [CrossRef]
36. Newman, J.C., Jr.; Vizzini, A.; Yamada, Y. *Fatigue Crack Growth Databases and Analyses for Threshold Behavior in Rotorcraft Materials*; DOT/FAA/AR-10/3; Department of Transportation: Washington, DC, USA, 2010.
37. Newman, J.C., Jr. Fracture analysis of surface- and through-cracked sheets and plates. *Eng. Fract. Mech.* **1973**, *5*, 667–689. [CrossRef]
38. Newman, J.C., Jr. A crack-opening stress equation for fatigue crack growth. *Int. J. Fract.* **1984**, *24*, R131–R135. [CrossRef]
39. Newman, J.C., Jr.; Kota, K.; Lacy, T.E. Fatigue and crack-growth behavior in a titanium alloy under constant-amplitude and spectrum loading. *Eng. Fract. Mech.* **2018**, *187*, 211–224. [CrossRef]
40. Newman, J.C.; Beukers, A. Effects of constraint on crack growth under aircraft spectrum loading. In *Fatigue of Aircraft Materials*; Beukers, A., deJong, T., Sinke, J., Vlot, A., Vogelesang, L.B., Eds.; Langley Research Center: Delft, The Netherlands, 1992; pp. 83–109.

41. Dugdale, D.S. Yielding of steel sheets containing slits. *J. Mech. Phys. Solids* **1960**, *8*, 100–104. [CrossRef]
42. Newman, J.C., Jr. *Fracture Mechanics Parameters for Small Fatigue Cracks*; ASTM STP 1149; American Society of Testing Materials: West Conshohocken, PA, USA, 1992; pp. 6–33.
43. Newman, J.C., Jr.; Swain, M.H.; Phillips, E.P. An assessment of the small-crack effect for 2024-T3. In *Small Fatigue Cracks*; Ritchie, R.O., Lankford, J., Eds.; The Metallurgical Society, Inc.: Warrendale, PA, USA, 1986; pp. 427–452.
44. Liu, J.Z.; Wu, X.R.; Ding, C.F.; Hu, B.R.; Wang, L.F.; Annigeri, B.; Vestergaard, L.H.; Romanowski, A.; Schneider, G.J.; Forth, S.C. Crack growth behavior and fatigue-life prediction based on worst-case near-threshold data of a large crack for 9310 steel. *Fatigue Fract. Eng. Mater. Struct.* **2002**, *25*, 467–480. [CrossRef]

Article

Computational Failure Analysis under Overloading

Slobodanka Boljanović [1],* **and Andrea Carpinteri** [2]

1   Mathematical Institute of the Serbian Academy of Sciences and Arts, 11000 Belgrade, Serbia
2   Department of Engineering and Architecture, University of Parma, 43124 Parma, Italy;
     andrea.carpinteri@unipr.it
*   Correspondence: slobodanka.boljanovic@mi.sanu.ac.rs; Tel.: +381-11-2630170

Abstract: The aim of this research work is to shed more light on performance-based design through a computational framework that assesses the residual strength of damaged plate-type configurations under overloading. Novel expressions are generated to analyze the power of crack-like stress raisers coupled with retardation effects. Analytical outcomes show that careful consideration of the overload location and crack size can be quite effective in improving safety design and failure mode estimation.

Keywords: analytical framework; fatigue crack; residual strength; retardation effect

Citation: Boljanović, S.; Carpinteri, A. Computational Failure Analysis under Overloading. *Metals* **2021**, *11*, 1509. https://doi.org/10.3390/met11101509

Academic Editor: George A. Pantazopoulos

Received: 2 August 2021
Accepted: 15 September 2021
Published: 23 September 2021

Publisher's Note: MDPI stays neutral with regard to jurisdictional claims in published maps and institutional affiliations.

1. Introduction

The well-known role of surface flaws in compromising bearing capacity during service operations, characterized by variable amplitude loading where overload and underload often exist, constantly warns that such stress raisers can cause sudden hazards in large moving systems.

An understanding of the load interaction effects on the fatigue crack extension, resulting from changes in the cyclic load level, is desirable to ensure the safety integrity and full functioning of structural components exposed to dynamic load environments with/without mixed loading modes, by employing reliable computational frameworks.

From the point of view of fracture mechanics, the driving mode interactions due to overload may be considered through several mechanisms such as residual stress [1], crack closure [2], crack tip blunting [3], strain hardening [4], crack branching [5] and reversed yielding [6].

The interest in the overload phenomenon, even if it has been researched for a long time, has become more pronounced in order to meet increasingly stringent damage tolerance-based demands for modern systems, combining relevant mechanisms and/or concepts. Thus, Elber [2] suggested that the concept of crack opening stress can be applied to generate the driving interactions in the vicinity of the crack tip due to overload. Budiansky and Hutchinson [7] employed the strip-yield hypothesis, generally attributed to the work of Dugdale [8] and Barenblatt [9], to model the plasticity-induced fatigue crack closure. Ohji et al. [10] introduced an incremental plasticity model incorporating kinematic hardening and crack growth simulation, extending the fatigue flaw in each stress cycle by a prescribed length that was equal to the finite element mesh size. Willenborg et al. [11] and Wheeler [12] suggested that residual compressive stresses can retard post-overload crack growth, and they developed their fracture mechanics models. Later, Suresh [5] discussed that retardation can persist even when the post-overload has traversed through the predicted zone of residual compressive stresses. Fleck [13] and Wang et al. [14] performed the finite element analysis to generate a plasticity-induced crack closure in two-dimensional configurations under plain-strain conditions.

Through a stability analysis under overload, Sander and Richard [15] employed the strip yield model and the modified generalized Willenborg model, implemented in the NASGRO [16] software package. Pavlou et al. [17] developed a fracture mechanics model based on strain-hardening fatigue mechanisms in the plastic zone in order to explore the influence of yield stress changes within the overload plastic zone on the fatigue crack growth rate. Huang et al. [18] introduced a modified Wheeler model incorporating an improved fatigue crack growth rate solution and the equivalent stress-intensity concept.

Further, Mohanty et al. [19] took into account Frost and Dugdale's fracture mechanics concept and the modified two-parameter crack growth model proposed by Jones et al. [20]. Harmain [21] proposed several modifications to Wheeler's crack growth idea involving an effective stress intensity factor based on Elber's crack closure concept [2], the relationship between overload ratio and the Wheeler's exponent, and the fatigue crack growth rate calculation. Boljanović and Maksimović [22] examined the fatigue performance of a through-crack under overloading using the Kujawski crack growth concept [23] together with the Wheeler retardation concept. Further, Boljanović et al. [24] analyzed the stability of the same stress raiser, linking the Zhan et al. concept [25] and the Wheeler concept [12] together with relevant solutions suggested by Richard et al. [26] for assessing the stress intensity factor under mixed mode loading with overload.

In the present work, the trends of fatigue strength degradation are evaluated taking into account the overload effect. In order to identify and characterize the potential features of such a phenomenon, an analytical framework is proposed, and a detailed body of evidence is provided for evaluating detrimental effects of a crack-like through flaw coupled with retardation effect due to overloading. Solutions from relevant literature demonstrate that this computational tool is able to provide a physical interpretation of the overload mechanism governing the propagation of the fatigue flaw so that it is possible to generate stress intensities and residual life through a consistent set of fracture-mechanics parameters.

2. Driving Mode Analysis under Cyclic Loading

Achieving high safety performances necessarily requires a detailed analysis of the fatigue-critical hot spots in large moving systems through relevant fracture mechanics-based crack growth concepts [19,24,27,28]. Hence, the driving mode progression due to crack-like edge flaw (Figure 1) is evaluated by employing the Huang–Moan crack growth concept [29], expressed as follows:

$$\frac{da}{dN} = C(M\,\Delta K)^m \tag{1}$$

where da/dN and a are the crack growth rate and crack length, respectively, ΔK is the stress intensity factor range, C and m are material parameters experimentally obtained, and N is the number of loading cycles.

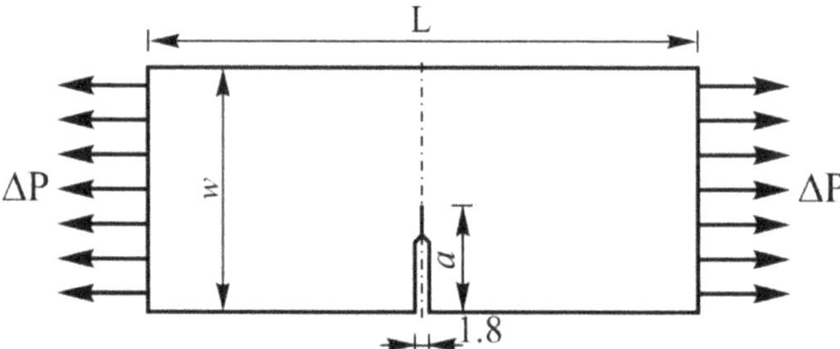

Figure 1. Geometry of the plate with edge crack-like flaw subjected to cyclic loading with overload (unit: mm).

Further, relevant interactions between in-service loading profiles and environmental effects are theoretically examined by means of the following fracture mechanics parameter [29]:

$$M = \begin{cases} (1 - R)^{-\beta_1} & -5 \leq R < 0 \\ (1 - R)^{-\beta} & 0 \leq R < 0.5 \\ \left(1.05 - 1.4R + 0.6R^2\right)^{-\beta} & 0.5 \leq R < 1 \end{cases} \tag{2}$$

where R is the stress ratio, β_1 and β are material parameters, experimentally obtained in the case of one negative and two positive cyclic loading domains [29], respectively.

Through a safety-relevant analysis herein presented, after integration crack growth rate (Equation (1)) from initial a_0 to final a_f crack length, the residual life can be evaluated as follows:

$$N = \int_{a_0}^{a_f} \frac{da}{C(M \, \Delta K)^m} \tag{3}$$

In order to make correct durability decisions and have well-coordinated fatigue responses to potential hazard events, a reliable assessment of residual bearing capacity and severity of plate-type system damage is needed. According to damage tolerance requirements, nonlinear driving force interactions in the vicinity of the crack tip are herein explored through the stress intensity range expressed as follows:

$$\Delta K = \frac{\Delta P}{w \, t} f_I\left(\frac{a}{w}\right) \sqrt{\pi \, a} \tag{4}$$

where ΔP is applied force range, w and t are width and thickness of the plate, respectively.

Further, in order to quantify the effect of an edge crack-like through flaw and geometry of considered plate, the following fracture mechanics-based parameter is employed [19].

$$f_I\left(\frac{a}{w}\right) = 1.12 - 0.231 \left(\frac{a}{w}\right) + 10.55 \left(\frac{a}{w}\right)^2 - 21.72 \left(\frac{a}{w}\right)^3 + 30.39 \left(\frac{a}{w}\right)^4 \tag{5}$$

3. Crack Growth Evolution Taking into Account the Retardation Effect

Variable amplitude cyclic-load environment may often have a significant impact on the strength capacity of large moving systems due to the appearance of overload, underload, and overload followed by underload, characterized by retardation, acceleration, and reduced retardation. Thus, if overload is applied, a relevant plastic zone of residual compressive stresses is generated in the vicinity of the crack tip. Such a nonlinear stress state field causes retardation in crack propagation by reducing the exiting crack tip driving force. In this context, fatigue analysis is herein performed employing the fracture mechanics-based concepts [29] combined with the retardation concept [12], i.e.,

$$\frac{da}{dN} = C_{pi}C(M\Delta K)^m \tag{6}$$

where C_{pi} is the fatigue parameter related to the overload effect.

Complex interactions in the zone where compressive stress states exist are theoretically examined using the following fracture mechanics solution proposed by Wheeler [12]:

$$C_{pi} = \begin{cases} \left(\dfrac{r_{pi}}{a_{ol} + r_{po} - a_i}\right)^p & ; \quad a_i + r_{pi} \leq a_{ol} + r_{po} \\ 1 & ; \quad a_i + r_{pi} \geq a_{ol} + r_{po} \end{cases} \tag{7}$$

where r_{pi} and a_i are the current plastic zone in the ith cyclic and corresponding crack size, and r_{po} and a_{ol} represent the overload plastic zone size and appropriate overload crack size, respectively, and p is the retardation exponent experimentally obtained.

Since the power of the overload effect decreases when the current plastic zone approaches the overload plastic zone, the relevant sizes of current plastic zone and monotonic overload plastic zone are computed [30,31] as follows:

$$r_{pi} = \frac{1}{\pi}\left(\frac{\Delta K}{2\,\sigma_{ys}}\right)^2 \quad r_{po} = \frac{1}{\pi}\left(\frac{K_{ol}}{\sigma_{ys}}\right)^2 \tag{8}$$

where ΔK and K_{ol} are the relevant stress intensity factor range and the stress intensity factor generated by an overload, respectively, and σ_{ys} is the yield strength of the material.

Furthermore, fatigue-based overload evaluations are herein performed in terms of the number of loading cycles, integrating the proposed solution for crack growth rate in Equation (6), i.e.,

$$N = \int_{a_0}^{a_f} \frac{da}{C_{pi}C(\Delta K)^m} \tag{9}$$

where a_0 and a_f are initial and final crack length, respectively.

With the need to reduce design life cyclic time and costs with ever more complex large moving systems, the possibility of assessing the failure strength against fatigue-induced loading interactions through advanced computational strategies is attractive. Thus, a damage tolerance-based analytical framework is herein developed, in which Euler's algorithm is implemented for generating fatigue resistance under overloading, as is discussed through several case study applications in the next section.

4. Residual Strength Design Using Developed Computational Framework

4.1. Fatigue Evaluations under Cyclic Loading with Overload

The failure performance design presented here tackles the crack-like edge stress raiser (Figure 1) under cyclic loading with overload. Through the life evaluations for the plate made of 6061 T6 aluminum alloy, the following material and loading parameters are employed: $C = 6 \times 10^{-11}$, $m = 3.2$, $\beta = 0.7$, $\lambda = 1$, $p = 0.21$, $\sigma_{ys} = 296$ MPa, $P_{max} = 11{,}772$ N with $R = 0$. Relevant geometrical sizes are characterized by $w = 50$ mm, $t = 3$ mm, $L = 180$ mm, and an initial crack length equal to $a_0 = 6$ mm is examined. By adopting that the crack length of a single overload is equal to $a_{ol} = 7.5$ mm, driving mode interactions are theoretically examined in the case of three overload stress ratios ($R_{ol} = 1.42$, 1.67, and 1.88) which are shown in Table 1.

Table 1. Evaluated number of loading cycles and applied force ranges with corresponding stress ratios under overloading.

R_{ol}	ΔP_{ol} (N)	$N^{exp.}$ (Cycles) [32]	$N^{cal.}$ (Cycles)
1.42	16,677	31,250	29,620
1.67	19,620	38,513	32,600
1.88	22,072	57,140	50,340

The fatigue vulnerability analysis performed via a novel computational framework evaluates the driving mode caused by through-crack and the residual life through Equations (1)–(5) coupled with Equations (6)–(9), respectively. Safety outcomes shown in Figure 2a generate the stress intensities in the vicinity of the crack tip caused by overload ($R_{ol} = 1.42$), whereas the number of loading cycles versus crack length is examined in Figures 2b and 3, and Table 1 in the case of three different overload stress ratios ($R_{ol} = 1.42$, 1.67, and 1.88). Further experimental outcomes discussed by Kumar [32] are employed to estimate the predictive capability of generated plate lives.

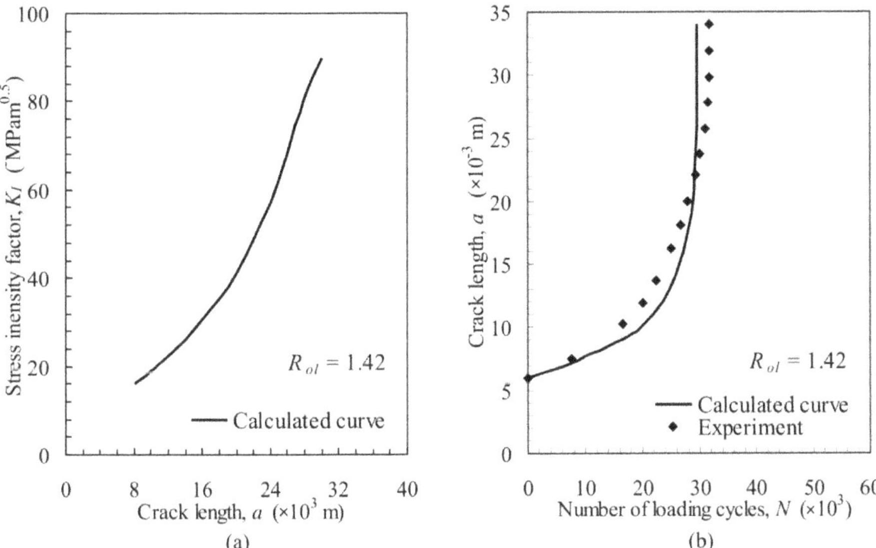

Figure 2. Fatigue resistance analysis: (**a**) K_I versus a, $R_{ol} = 1.42$ and (**b**) a versus N, $R_{ol} = 1.42$, calculated curves from the present work, experiments reported by Kumar [32].

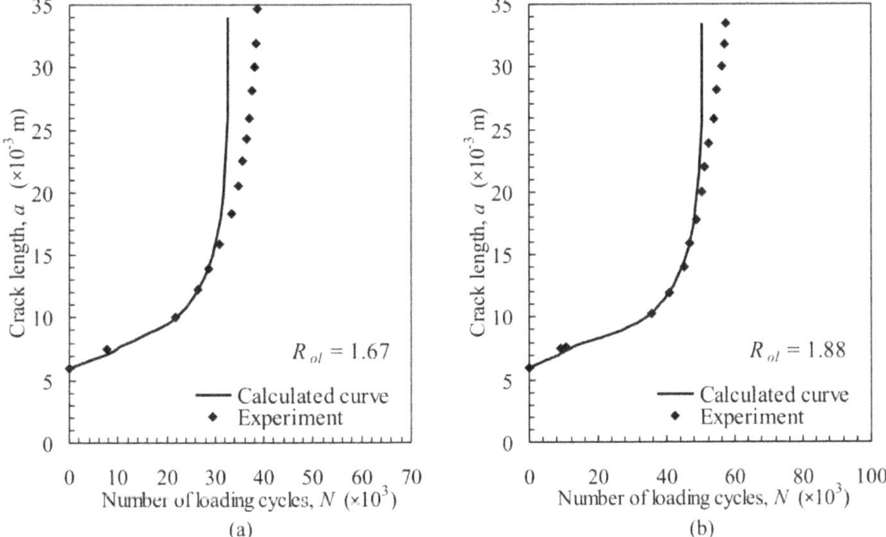

Figure 3. Fatigue resistance analysis: (**a**) a versus N, $R_{ol} = 1.67$ and (**b**) a versus N, $R_{ol} = 1.88$, calculated curves from the present work, experiments reported by Kumar [32].

Through different comparisons presented in this section, it can be inferred that the developed computational framework provides conservative estimates for plates with edge through-crack subjected to overload. In addition, the main contribution of the fatigue resistance analysis presented is that it can be stated that, for the considered plate with through-crack (characterized by $t = 3$ mm), the higher the overload ratio and applied overload force range, the more pronounced the impact of retardation effects under cyclic loading (see Table 1). Since complex interactions between the effects of through-crack and the stress ratio effect coupled with the effects of thin plate thickness are adequately evaluated via fatigue life (with respect to experiments [32]), the computational strategy herein proposed can significantly contribute to further improvements in safety design and optimization of large moving systems.

4.2. The Effect of Overload Crack Length and the Stress Ratio Effect on the Fatigue Strength

In this section, the strength performance under cyclic loading with single overload is designed for plates made of 6061 T6 aluminum alloy ($w = 50$ mm, $t = 4$ mm, $L = 200$ mm, Figure 1). Fatigue stability is explored in the case of overload crack length equal to $a_{ol} = 8.5$ mm, 14.5 mm, and 24.5 mm assuming the following overload and cyclic loading parameters: $P_{maxol} = 22{,}150$ N, $R_{ol} = 1.75$, $P_{max} = 15{,}500$ N, $R = 0.2$. The initial crack length is characterized by $a_0 = 5.5$ mm.

Damage tolerance-based assessments in terms of the residual life generated via novel analytical solutions (Equations (2)–(5)) are shown in Figure 4a for the three different single-overload conditions considered. Furthermore, the interactions between the effect of through-crack and overload effect are theoretically examined for plates ($w = 50$ mm, $t = 3.5$ mm, $a_0 = 4.5$ mm) subjected to three different values of stress ratio, i.e., $R = 0.1, 0.3$, and 0.5, respectively, as is shown in Figure 4b. Note that relevant force range and maximum overload force are equal to $P_{max} = 17{,}200$ N, $P_{maxol} = 24{,}420$ N, $R_{ol} = 1.95$, and $a_{ol} = 10.5$ mm, whereas material parameters are the same as those adopted in the previous section.

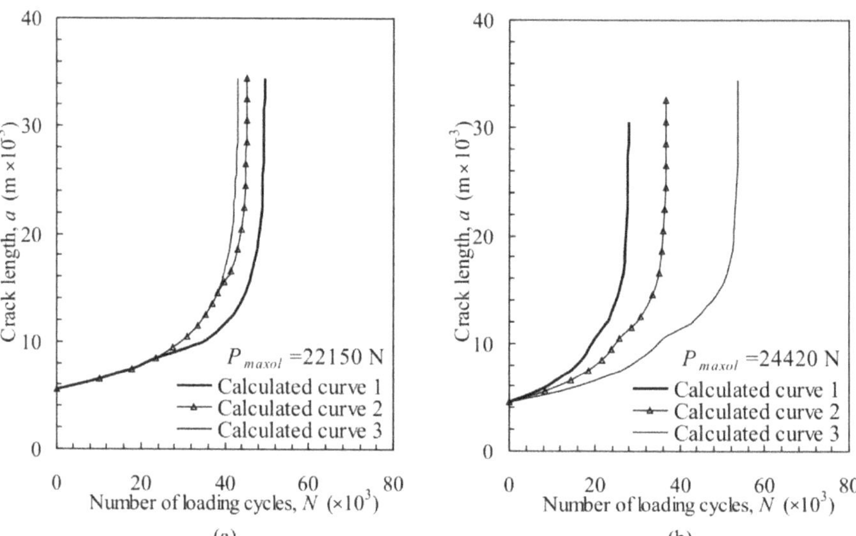

Figure 4. Fatigue resistance analysis: (**a**) *a* versus N ($1 - a_{ol} = 8.5$ mm, $2 - a_{ol} = 14.5$ mm, $3 - a_{ol} = 24.5$ mm) and (**b**) *a* versus N ($1 - R = 0.1$, $2 - R = 0.3$, $3 - R = 0.5$), calculated curves from the present work.

By analyzing relevant theoretical outcomes shown in Figure 4, it can be concluded that the number of loading cycles decreased from 49,450 to 45,180 due to the increase in overload crack length from 8.5 mm to 14.5 mm. Furthermore, if the stress ratio increases from 0.1 to 0.3, the number of loading cycles increases from 27,820 to 36,790. Evidently, an increase in overload crack length leads to a decrease in the retardation effect due to overloading. In addition, increasing the value of stress ratio under cyclic loading can contribute to increasing the impact of overload leading to an additional increase in residual life to failure.

4.3. Fatigue Evaluations under Mixed Mode Loading with Overload

Finally, the performance design carries out the driving mode progression due to the mixed mode flaws (Figure 5). The fatigue life is assessed here in the case of plate made of 2024 T3 aluminum alloy ($w = 52$ mm, $t = 6.5$ mm $\sigma_{ys} = 324$ MPa, $E = 73.1$ GPa, $\beta = 0.7$, $p = 0.21$, $C = 5 \times 10^{-11}$, $m = 3.05$). Further, maximum force $P_{max} = 7197$ N (with stress ratio $R = 0.1$) and relevant maximum overload force $P_{maxol} = 17{,}993$ N (with overload stress ratio $R_{ol} = 0.1$ applied at crack length $a_{ol} = 20.4$ mm) are adopted in order to analyze the loading angle effect of crack-like flaw, whose initial length is equal to $a_0 = 17.75$ mm.

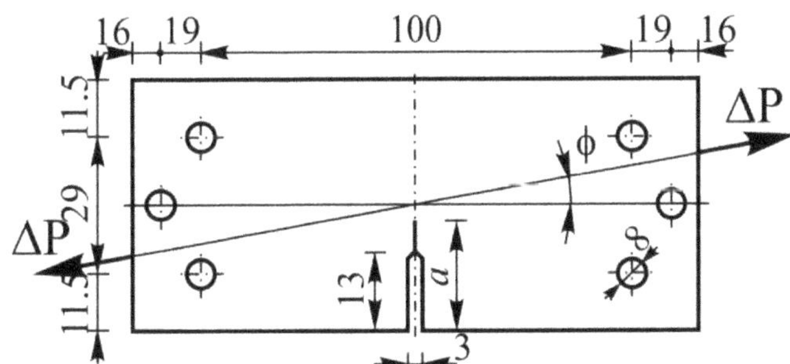

Figure 5. Geometry of the plate with edge crack-like flaw subjected to mixed-mode loading with overload. All dimensions are in millimeters.

In-service large moving systems often face a complex fatigue environment which can cause the formation of mixed-mode and/or multi-axial flaws. From the point of view of fracture mechanics, the detrimental effects of such randomly oriented quantities, represented as manufacturing and in-service flaws/defects, should be analyzed through the equivalent stress intensity factor [33]. Thus, the stress state in the vicinity of the crack tip, where combined loading modes exist (see Figure 5), may be evaluated as follows:

$$\Delta K_{eq} = \left(\Delta K_I^4 + 8\Delta K_{II}^4 \right)^{0.25} \tag{10}$$

Fatigue-induced interactions caused by local mode I and mode II load environment are herein assessed via relevant stress intensity factors [34], i.e.,

$$\Delta K_I = \frac{\Delta P}{w\, t} \cos\phi\, f_I\left(\frac{a}{w}\right) \sqrt{\pi a} \tag{11}$$

$$\Delta K_{II} = \frac{\Delta P}{w\, t} \sin\phi\, f_I\left(\frac{a}{w}\right) \sqrt{\pi a} \tag{12}$$

where ΔP is applied force ranges, a represents the crack length, and w and t are the width and thickness of the plate, respectively.

Through the failure resistance design, the crack growth rate and residual life are evaluated employing Equations (1)–(3) and Equations (6)–(8) together with Equations (9)–(12) for three different loading profiles i.e., $\phi = 18°$, $36°$, and $54°$, respectively. Figure 6a presents the equivalent stress intensity factors evaluated in the case of three different loading angles, whereas the residual strength in terms of the number of loading cycles is plotted in Figures 6b and 7a,b.

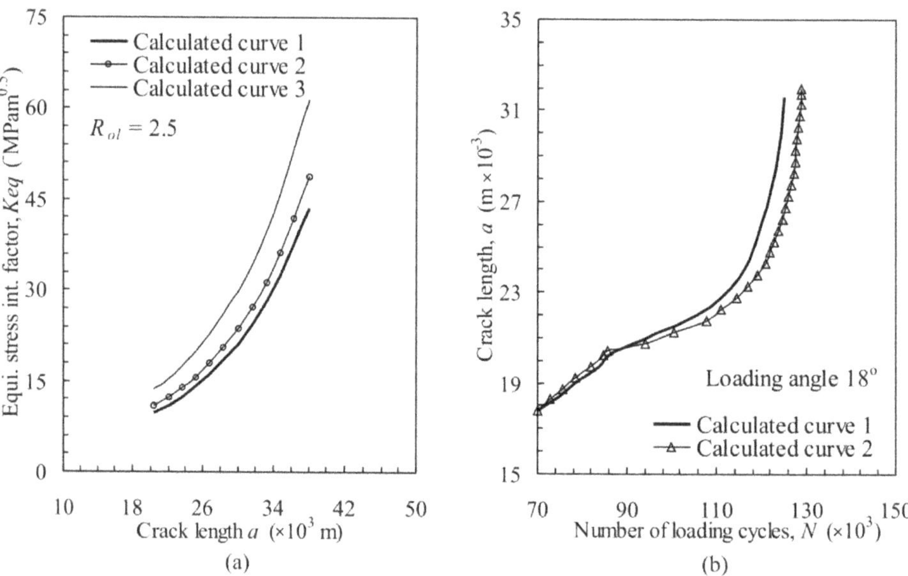

Figure 6. Fatigue resistance analysis: (**a**) K_{eq} versus a ($1 - \phi = 18°$, $2 - \phi = 36°$, $3 - \phi = 54°$), calculated curves from the present work, and (**b**) a versus N ($\phi = 18°$), 1—calculated curve from the present work, 2—calculated curve reported by Mohanty et al. [19].

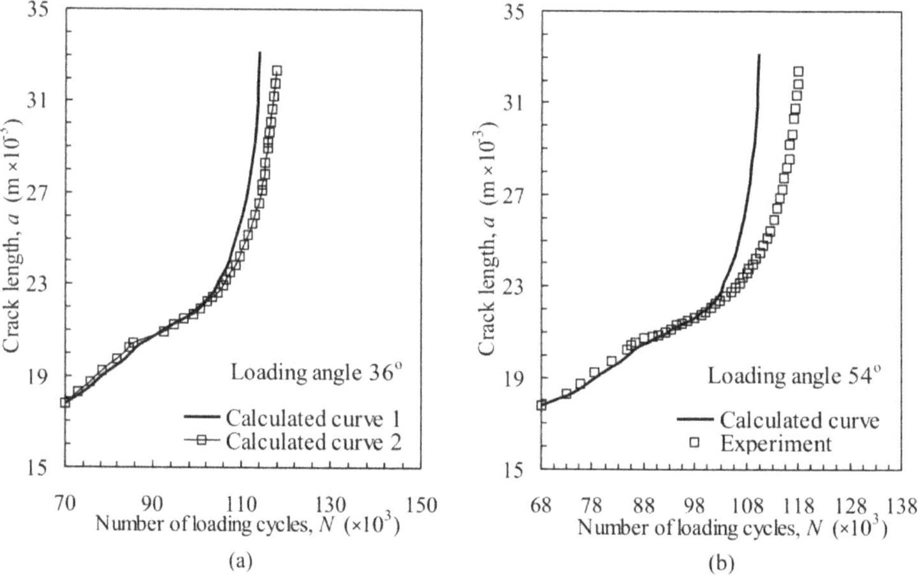

Figure 7. Fatigue resistance analysis: (**a**) a versus N ($\phi = 36°$), 1—calculated curve from the present work, 2—calculated curve from Reference [19] and (**b**) a versus N ($\phi = 54°$), calculated curve from the present work, experiments reported by Mohanty et al. [19].

Further, literature-based experimental/theoretical outcomes reported by Mohanty et al. [19] are examined for the same mixed mode configurations in order to assess the predictive capability of the generated computational framework. From Figures 6 and 7, it can be inferred that relevant experimental data and theoretical outcomes are in quite good agreement. Further, fatigue lives evaluated through this research work are compared with those discussed by Boljanović et al. [24] and Mohanty et al. [19], as is shown in Table 2. Note that relevant calculations from [24] represent those in which stress intensities due to the fact that mixed modes were analyzed using the Richard et al. [26] concept.

Table 2. Evaluated number of loading cycles and loading angles for relevant mixed mode conditions.

ϕ (°)	$N^{\,exp./cal.}$ (Cycles) [19]	$N^{\,cal.}$ (Cycles) [24]	$N^{\,cal.}$ (Cycles)
18	128,660	123,100	125,100
36	117,890	114,100	114,200
54	118,080	109,700	110,400

Comparisons in Table 2 indicate that relevant concepts discussed through this paper and those proposed within [24] adequately generate interactions between the stress raiser effects and mixed mode effects under overloading, where the framework previously developed provides a more conservative trend of estimates with respect to fatigue outcomes reported in [19]. It is evident that theoretical outcomes represent high performance/quality estimates.

Further, it should be noted that, during mandatory large moving systems inspections/controls in the case of loading angles $\phi = 36°$ and $54°$, a computational framework herein developed can have an important role since it generates residual life values closer to relevant experimental/theoretical outcomes [19] than the ones discussed in [24].

Moreover, the safety analysis demonstrates that retardation effect due to overloading is pronounced for loading angle $\phi = 18°$, while its impact in the case of $\phi = 36°$ and $54°$ decreased by about 10% if the edge crack-like flaw is subjected to mixed (I and II) loading modes.

5. Conclusions

Due to heavy usage, large moving systems usually work and deteriorate in a variable amplitude dynamic environment, where deleterious factors caused by localized flaws can seriously compromise their bearing capacities. Thus, reliable and target-oriented assessments of fatigue degradation under overload, according to damage tolerance requirements, are very important. A challenging and also interesting task is to monitor/control the failure strength through the residual life evaluations simultaneously and independently in the case of axial loading and mixed mode loading, taking into account the load interaction effect. In this context, the present research work proposes a novel analytical framework for analyzing interactions between the effects of crack-like through flaws and overload effect.

In order to explore crack growth retardation, a two-parameter driving force concept is combined with the fracture mechanics concept proposed by Wheeler. Further, relevant fatigue life solutions are established which are essential to the reliable control of through-flaw configurations under axial and mixed mode loading. Several case study applications are given to demonstrate the fatigue assessments under targeted cyclic load profiles and to verify the developed computational tool taking into account the effect of overload stress ratio and the loading angle effects due to mixed modes. In addition, a significant contribution of this research is that it supplies notable information about overload mechanisms and sheds a light on improving the safe-integrity performance design of failure-critical aeronautical systems.

Author Contributions: Conceptualizations, S.B. and A.C.; methodology, S.B. and A.C.; software, S.B.; validation, S.B. and A.C. writing-original draft preparation, S.B.; writing-review and editing S.B. and A.C. All authors have read and agreed to the published version of the manuscript.

Funding: This research received no external funding.

Data Availability Statement: Not available.

Acknowledgments: The present scientific research was supported by the Serbian Ministry of Education, Science and Technological Development through the Mathematical Institute of the Serbian Academy of Sciences and Arts, Belgrade and the COST Association, Brussels, Belgium within the Action CA 18203, which is gratefully acknowledge.

Conflicts of Interest: The authors declare no conflict of interest.

References

1. Schijve, J.; Broek, D. Crack propagation: The results of a test programme based on a gust spectrum with variable amplitude loading. *Aircraft Eng. Aerospace Techn.* **1962**, *34*, 314–316. [CrossRef]
2. Elber, W. The significance of fatigue crack closure. In *Damage Tolerance in Aircraft Structures*; ASTM STP: Philadelphia, PA, USA, 1971; Volume 486, pp. 230–242.
3. Christensen, R.H. Fatigue crack. In *Fatigue Cracking, Fatigue Damage and Their Detection, Metal Fatigue*; Sines, G., Waisman, J.L., Eds.; McGraw-Hill Book Co.: New York, NY, USA, 1959; pp. 376–412.
4. Jones, R.E. Fatigue crack growth retardation after single-cycle peak overload in Ti-6Al-4V titanium alloy. *Eng. Fract. Mech.* **1973**, *5*, 585–604. [CrossRef]
5. Suresh, S. Micromechanisms of fatigue crack growth retardation following overloads. *Eng. Fract. Mech.* **1983**, *18*, 577–593. [CrossRef]
6. Nicoletto, W. Fatigue crack-tip field measurements. In *Nonlinear Fracture Mechanics*; ASTM STP: Philadelphia, PA, USA, 1989; Volume 995, pp. 415–432.
7. Budiansky, B.; Hutchinson, J.W. Analysis of closure in fatigue crack growth. *ASME J. Appl. Mech.* **1978**, *45*, 267–276. [CrossRef]
8. Dugdale, D.S. Yielding of steel sheets containing slits. *J. Mech. Phys. Solids* **1960**, *8*, 100–104. [CrossRef]
9. Barenblatt, G.I. The mathematical theory of equilibrium cracks in brittle fracture. In *Advances in Applied Mechanics 7*; USSR Academy of Sciences: Moscow, Russia, 1962; pp. 55–129.
10. Ohji, K.; Ogura, K.; Yoshiji, O. Cyclic analysis of a propagating crack and its correlation with fatigue crack growth. *Eng. Fract. Mech.* **1975**, *7*, 457–464. [CrossRef]
11. Willenborg, J.D.; Engle, R.M.; Wood, H.A. *A Crack Growth Retardation Model Using an Effective Stress Concept*; Report AFFDL-TM-71-1-FBR; Air Force Flight Laboratory, Wright-Patterson Air Force Base: Dayton, OH, USA, 1971.
12. Wheeler, O.E. Spectrum loading and crack growth. *J. Bas. Eng. Trans. ASME Ser. D* **1972**, *94*, 181–186. [CrossRef]
13. Fleck, N.A. Finite element analysis of plasticity induced crack closure under plane strain conditions. *Eng. Fract. Mech.* **1986**, *25*, 441–449. [CrossRef]
14. Wang, C.H.; Rose, L.R.F.; Newman, J.C. Closure of plane-strain cracks under large-scale yielding conditions. *Fatigue Fract. Eng. Mater. Struct.* **2002**, *25*, 127–139. [CrossRef]
15. Sander, M.; Richard, H.S. Lifetime predictions for real loading situations–concepts and experimental results of fatigue crack growth. *Int. J Fatigue* **2003**, *25*, 999–1005. [CrossRef]
16. NASGRO. *Fatigue Crack Growth Computer Program "NASGRO"*; Version 3.0-Reference Manual, JSC-22267B; NASA, Lyndon B. Johnson Space Center: Houston, TX, USA, 2000.
17. Pavlou, D.G.; Vlachakis, N.V.; Pavlou, M.G.; Vlachakis, V.N. Estimation of fatigue crack growth retardation due to crack branching. *Comp. Mater. Sci.* **2004**, *29*, 446–452. [CrossRef]
18. Huang, X.; Moan, T.; Weicheng, C. An engineering model of fatigue crack growth under variable amplitude loading. *Int. J. Fatigue* **2008**, *30*, 2–10. [CrossRef]
19. Mohanty, J.R.; Verma, B.B.; Ray, P.K. Prediction of fatigue life with interspersed mode-I and mixed-mode (I and II) overloads by an exponential model: Extensions and improvements. *Eng. Fract. Mech.* **2009**, *76*, 454–468. [CrossRef]
20. Jones, R.; Molent, L.; Pitt, S. Crack growth of physically small cracks. *Int. J. Fatigue* **2007**, *29*, 1658–1667. [CrossRef]
21. Harmain, G.A. A model for predicting the retardation effect following a single overload. *Theor. Appl. Fract. Mech.* **2010**, *53*, 80–88. [CrossRef]
22. Boljanović, S.; Maksimović, S. Computational mixed mode failure analysis under fatigue loadings with constant amplitude and overload. *Eng. Fract. Mech.* **2017**, *174*, 168–179. [CrossRef]
23. Kujawski, D. A new $(\Delta K^{+}K_{max})^{0.5}$ driving force parameter for crack growth in aluminium alloy. *Int. J. Fatigue* **2001**, *23*, 733–740. [CrossRef]
24. Boljanović, S.; Maksimović, S.; Carpinteri, A. Fatigue-resistance evaluations for mixed mode damages under constant amplitude and overload. *Theor. Appl. Fract. Mech.* **2020**, *108*, 102599. [CrossRef]
25. Zhan, W.; Lu, N.; Zhang, C. A new approximate model for the R-ratio effect on fatigue crack growth rate. *Eng. Fract. Mech.* **2014**, *119*, 85–96. [CrossRef]
26. Richard, H.A.; Linning, W.; Henn, K. Fatigue crack propagation under combined loading. *Forensic Eng.* **1991**, *3*, 99–109.
27. Carpinteri, A.; Spagnoli, A.; Vantadori, S.; Bagni, C. Structural integrity assessment of metallic components under multiaxial fatigue: The C-S criterion and its evaluation. *Fatigue Fract. Eng. Mater. Struct.* **2013**, *36*, 870–883. [CrossRef]
28. Carpinteri, A.; Spagnoli, A.; Ronchei, C.; Scorza, D.; Vantadori, S. Critical plane criterion for fatigue life calculation: Time and frequency domain formulations. *Procedia Eng.* **2015**, *101*, 518–523. [CrossRef]
29. Huang, X.; Moan, T. Improved modeling of the effect of R-ratio on crack growth rate. *Int. J. Fatigue* **2007**, *29*, 591–602. [CrossRef]

30. Antunes, F.V.; Borrego, L.F.P.; Costa, J.M.; Ferreira, J.M. A numerical study of fatigue crack closure induced by plasticity. *Fatigue Fract. Eng. Mater. Struct.* **2004**, *27*, 825–835. [CrossRef]
31. Rushton, P.A.; Taheri, F. Prediction of crack growth in 350WT steel subjected to constant amplitude with over- and under-loads using a modified wheeler approach. *Mar. Struct.* **2003**, *16*, 517–539. [CrossRef]
32. Kumar, R. Prediction of delay cycles due to instant of single overload cycles. *Eng. Fract. Mech.* **1992**, *42*, 563–571. [CrossRef]
33. Tanaka, K. Fatigue crack propagation from a crack inclined to the cyclic tensile axis. *Eng. Fract. Mech.* **1974**, *6*, 493–507. [CrossRef]
34. Chao, Y.J.; Liu, S. On the failure of cracks under mixed-mode loads. *Int. J. Fract.* **1997**, *87*, 201–223. [CrossRef]

Article

Elastoplastic Fracture Analysis of the P91 Steel Welded Joint under Repair Welding Thermal Shock Based on XFEM

Kai Yang [1,2], Yingjie Zhang [1,2] and Jianping Zhao [1,2,*]

[1] School of Mechanical and Power Engineering, Nanjing Tech University, No. 30 Puzhu South Road, Pukou District, Nanjing 211816, China; kyang95@163.com (K.Y.); yjzhang_njtech@163.com (Y.Z.)
[2] Jiangsu Key Lab of Design and Manufacture of Extreme Pressure Equipment, Nanjing Tech University, No. 30 Puzhu South Road, Pukou District, Nanjing 211816, China
* Correspondence: jpzhao@njtech.edu.cn; Tel.: +86-25-5813-9951

Received: 29 August 2020; Accepted: 22 September 2020; Published: 25 September 2020

Abstract: P91 steel is a typical steel used in the manufacture of boilers in ultra-supercritical power plants and heat exchangers in nuclear power plants. For the long-term serviced P91 steel pressurized structures, the main failure mode is the welded joint failure, especially the heat affected zone (HAZ) failure. Repair welding technique is an effective method for repairing such local defects. However, the thermal shock composed of high temperature and thermal stress in the repair welding process will pose a critical loading condition for the existing defects near the heat source which cannot be detected by conventional means. So, the evaluation of structural integrity for the welded joint in the thermal-mechanical coupling field is necessary. In this work, the crack propagation law in the HAZ for the P91 steel welded joint was investigated under repair welding thermal loads. The weld repair model of the P91 steel welded joint was established by ABAQUS. The transient temperature field and stress field in repair welding process were calculated by relevant user subroutines and sequential coupling simulation method. The residual stress was determined by the impact indentation strain method to verify the feasibility of the finite element (FE) model and simulation method. In order to obtain the crack propagation path, the elastoplastic fracture analysis of the welded joint with initial crack was performed based on the extended finite element method (XFEM). The influence of different welding linear energy on the crack propagation was analyzed. The results show that the cracks in the HAZ propagate perpendicular to the surface and tend to deflect to the welding seam under repair welding thermal loads. The crack propagation occurs in the early stage of cooling. Higher welding linear energy leads to larger HAZ and higher overall temperature. With the increase of welding linear energy, the length and critical distance of the crack propagation increase. Therefore, low welding linear energy can effectively inhibit the crack propagation in the HAZ. The above calculation and analysis provide a reference for the thermal shock damage analysis of repair welding process, which is of great significance to improving the safety and reliability of weld repaired components.

Keywords: fracture analysis; welded joint; repair welding thermal shock; XFEM; welding linear energy

---

1. Introduction

P91 steel is widely used in boiler components of ultra-supercritical power plants due to its excellent creep strength and steam corrosion resistance at high temperature [1]. It is also one of the common choices of steel pipe material for heat exchangers in nuclear power plants [2]. In practical application, P91 steel is usually connected by welding to form the pressure equipment components. During long-term service, local defects will occur inevitably in pressure equipment due to the effect of harsh service environment or improper operation procedure, especially in vulnerable parts, such as welded joints [3,4]. As a Cr-Mo heat-resistant steel, the main failure mode of the P91 steel welded joints is the type IV cracking near the fine-grained zone (FGHAZ) (shown in Figure 1), which will result in the degradation of structural safety [5]. An efficient and economic method to repair the cracking region is to remove the area with local defects and then refill it with welding rods by repair welding technique [6,7], which greatly extends the service life of pressure equipment. However, conventional defect detection means, including radiographic testing, ultrasonic testing, magnetic particle testing, penetrant testing and eddy current testing, have a corresponding detection sensitivity. This will result in undetected defects existing in the structure.

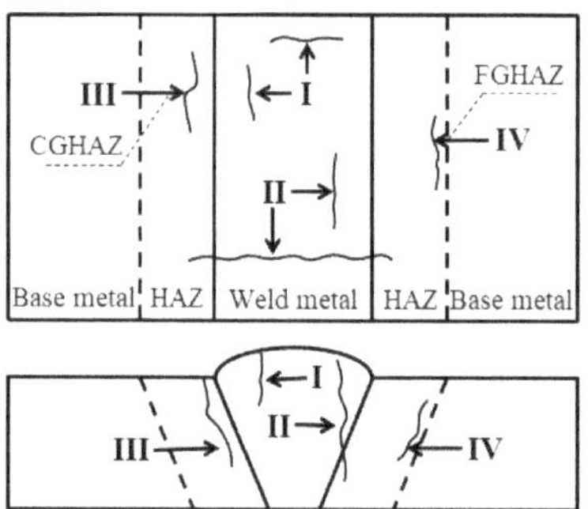

Figure 1. Crack classification of Cr-Mo steel welded joint, reproduced from [5], with permission from International Materials Reviews, 2020.

Repair welding is a rapid welding heat transfer process, and the arc temperature during welding process may reach thousands of degrees centigrade [8]. The welding heat source loads a large amount of heat flow into the welding seam and nearby area in a short time. The high temperature and the unsteady thermal stress act on the weldment in the form of thermal shock. On the one hand, it will cause the formation of hot cracks in the welding seam; on the other hand, the interaction between existing cracks and thermal-mechanical coupling field will lead to the crack propagation and even the failure of the whole component. In order to ameliorate the quality of weldments and weld repaired components, it is necessary to study the effect of the welding process on the cracking behavior. Alvarez et al. [9] compared the hot cracking sensitivity of tungsten inert gas welding (TIG) and laser beam welding (LBW) by analyzing the microstructure and chemical composition of the welding seam of 718 alloy. Chelladurai et al. [10] studied the solidification cracking behavior of 316 stainless steel under different energy transfer modes by adjusting the pulse parameters in pulsed laser welding (PLW). Hosseini et al. [11] investigated the effect of heat input and welding speed in electron beam welding (EBW) on the hot cracking sensitivity of AA2024-T351 alloy experimentally. Coniglio et al. [12] explored the initiation mechanism of solidification crack in the arc welding seam for 6060 aluminum alloy theoretically and numerically. Wei et al. [13] developed a software which can automatically plot the driving force and resistance curves of solidification cracks according to the numerical and experimental results. Bordreuil et al. [14] established a solidification cracking model which can predict microstructure and pore nucleation in the welding seam by combining cellular automata model with intergranular fluid flow model. Agarwal et al. [15] analyzed the influence of metallurgical factors and molten pool shape on the solidification cracking in laser beam welding of advanced high strength steels numerically and experimentally. Jiang et al. [16] researched the effect of repair width on residual stress for the composite plate to reduce the probability of interface cracking. Hyde et al. [17] studied the creep crack propagation behavior of P91 steel weldment under constant loads by both experimental method and finite element (FE) method. Pandey et al. [18] investigated experimentally the effect of different diffusible hydrogen concentrations in weld metal on hydrogen-induced cracking features of P91 steel. Zhang et al. [19] applied the high energy spark deposition (HESD) method to weld repair and tested the fracture properties of the repair welding seam. He et al. [20] performed the elastic fracture analysis of cracked aluminum alloy plate during metal inert gas welding (MIG) and cooling processes.

According to the existing literature, a lot of theoretical and experimental studies have been carried out on the hot cracks in welding processes. Almost all the current numerical simulations of cracking behavior for welding focus on the structure without crack or with stationary crack. The thermal effect of welding, especially repair welding, will make the existing cracks which are neglected by conventional detection methods further extend and even penetrate through the whole structure. However, there is little research on the crack propagation behavior under repair welding thermal shock. Therefore, it is necessary to analyze the structural integrity of weldments under repair welding thermal loads.

Combined with the specific damage model, the extended finite element method (XFEM) can simulate the ductile crack growth behavior accurately [21]. Belytschko [22] first proposed the embryonic form of the XEFM. It was independent of mesh generation and enhanced the finite element approximation by adding discontinuous enrichment functions to the displacement field near the crack tip. Moës et al [23] introduced the step function and crack tip asymptotic function to describe the crack surface and crack tip, respectively, making it successfully used in the analysis of fracture mechanics, and called this technique "extended finite element method". Daux et al. [24] added several different asymptotic functions and step functions to the crack tip and crack surface to simulate the crack branching. Chessa et al. [25] improved the convergence of the blending element in the XFEM (shown in Figure 2) by utilizing the extended strain method. Stolarska et al. [26] used the level set method to locate the crack location, which further improved the theory of the XFEM. For completing the modelling of dynamic crack, Song et al. [27] applied the phantom node method to the XFEM, which implemented the definition of the crack propagation behavior within the framework of the conventional finite element method. Until now, FE software such as ABAQUS, ANSYS and LS-DYNA have implemented the XFEM program into the function module of fracture analysis. This paper uses the XFEM to predict how the repair welding thermal shock affects the cracking behavior of the heat affected zone (HAZ) in a P91 steel welded joint. The influence of welding linear energy on the cracking features has been studied, which provides a reference for repair welding of structures containing defects. This study has guiding significance for the life extension of pressure equipment under long-term service or extended service.

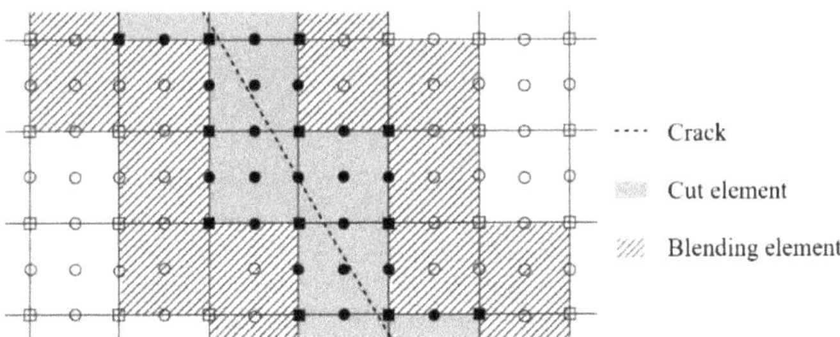

Figure 2. Cut element and blending element in the extended finite element method (XFEM).

2. FE Model and Simulation Method

2.1. Model Geometry

Two P91 steel plates with a size of 200 mm × 100 mm × 12 mm (length, width and thickness, respectively) are butt welded with four welding layers, including two layers of root welding (welding layers 1 to 2), one layer of filler welding (welding layer 3) and one layer of cover welding (welding layer 4). The selection of the form and size of welding groove is based on the criteria of Yang et al. [28]. A V-shaped groove is adopted, the root gap and blunt edge are both 2 mm, the groove angle is 60°. The area containing a type IV crack in the HAZ will be removed and then refilled by repair welding. The size of weld repair area is 80 mm × 8 mm × 3 mm (length, width and thickness, respectively), and the angle of groove surface for repair welding is 30°. Figure 3 shows the schematic drawing of the model geometric dimensions.

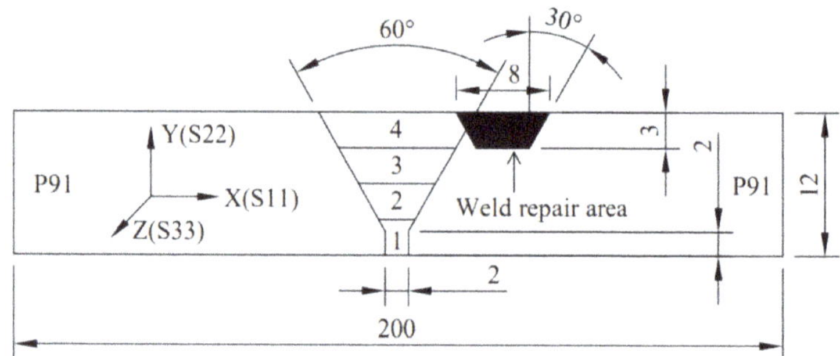

Figure 3. Schematic drawing of the model geometry. (unit: mm).

The finite element modelling tool of thermal-mechanical coupling process and fracture behavior is ABAQUS 6.14-5 (Dassault Systemes, Paris, France). In order to save the calculating work and avoid the convergence problem in 3-D computational fracture mechanics, a 2-D plane strain analysis model is set up. The plane studied in this paper is the middle cross section of the plate, as shown in Figure 4. Mesh refinement is carried out in the welding seam and HAZ, where exist high heat flux loading, to overcome the mesh-sensitivity problem. The area far away from the welding seam and HAZ is coarsened appropriately to reduce the calculation time. The transitional mesh is adopted between the refined mesh and the coarsened mesh. The FE model and meshing are shown in Figure 5. A total of 9063 elements and 9322 nodes are meshed. The element edge size for the refined mesh is 0.25 mm, and that for the coarsened mesh is 2 mm.

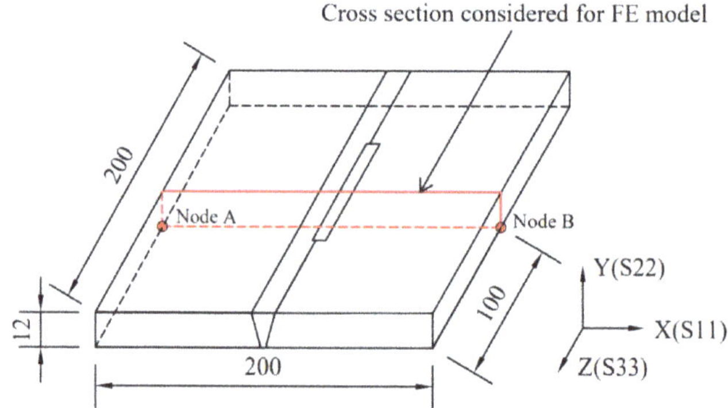

Figure 4. The location of the cross section for analysis. (unit: mm).

The sequential coupling method is used in the thermal-mechanical coupling simulation. Firstly, the thermal analysis is performed, and then the temperature field results are applied to the nodes of the mechanical model to carry out the mechanical analysis. The removal and deposition of the metal are implemented by the method of Birth and Death Element. A 4-node linear heat transfer quadrilateral (DC2D4) element is adopted for the thermal analysis, and a 4-node bilinear plane strain quadrilateral reduced-integration (CPE4R) element is adopted for the mechanical analysis. The topological relationship for meshing between the thermal model and the mechanical model is consistent.

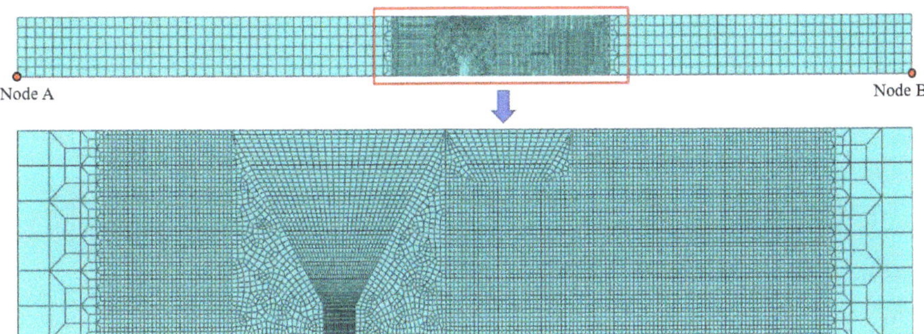

Figure 5. Finite element (FE) model and meshing.

2.2. Material Properties and Heat Source Model

2.2.1. Material Property

Temperature-dependent nonlinear features are considered for the thermal physical properties and mechanical properties of P91 steel, as shown in Figure 6. The thermal physical properties include thermal conductivity, specific heat and thermal expansion coefficient, while the mechanical properties include Young's modulus, Poisson's ratio and yield strength. The density of P91 steel is assumed to be constant at a value of 7780 kg/m^3. The material properties of the weld metal are also considered in the FE simulation for obtaining higher accuracy. The solidus and liquidus temperatures of both base metal and weld metal are set at 1420 °C and 1500 °C, respectively. The melting heat and solidification heat of the material are both set at 260 kJ/kg.

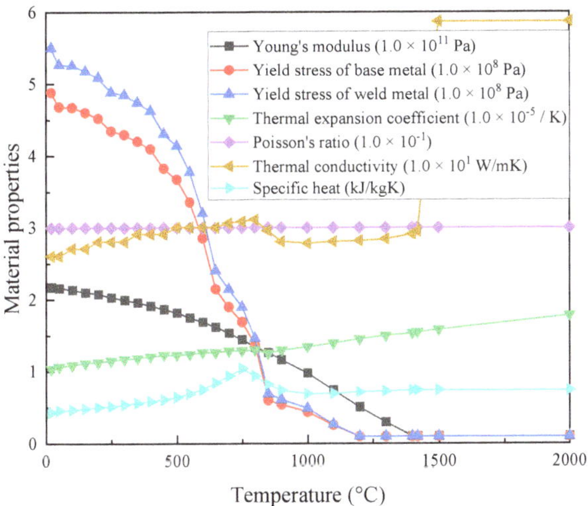

Figure 6. Temperature-dependent thermal physical properties and mechanical properties, for base metal and weld metal, used in the FE simulation, data from [29].

2.2.2. Heat Source Definition

The heat source loads should reflect the actual temperature change during welding as accurately as possible. The distributed heat flux (DFLUX) has been widely used because of its mature theory. Among the distributed heat flux, Gaussian heat flux distribution, uniform body heat flux distribution and double ellipsoidal heat flux distribution are commonly used [30].

The thermal exchange during welding is mainly the thermal conduction inside the weldment, which follows Fourier's law (Equation (1)), and the governing equation of temperature field (Equation (2)) follows the law of conservation of energy.

$$R = -\lambda \times \frac{\partial T}{\partial x} \tag{1}$$

$$\rho c \times \frac{\partial T}{\partial t} = \frac{\partial}{\partial x} \times \left(\lambda \times \frac{\partial T}{\partial x}\right) + \frac{\partial}{\partial y} \times \left(\lambda \times \frac{\partial T}{\partial y}\right) + \frac{\partial}{\partial z} \times \left(\lambda \times \frac{\partial T}{\partial z}\right) + Q \tag{2}$$

where:

R is the heat flux (W/m^2);

λ is the thermal conductivity (W/m·k);

T is the distribution function of temperature field (K);

ρ is the density (kg/m^3);

c is the specific heat (J/kg·K);

t is the transient time (s);

Q is the intensity of thermal energy (W), including the thermal energy generated by the heat source and the thermal energy generated by the solid-liquid phase change.

The conventional uniform body heat flux distribution is defined by Equation (3) [29].

$$q = \frac{\eta \times U \times I}{V_{act}} \tag{3}$$

where q is the heat generation rate by heat source; η is the arc efficiency factor; U is the arc voltage; I is the welding current; V_{act} is the action volume of the heat source.

The conventional uniform body heat flux distribution adopts amplitude curves to control the piecewise changes of the heat generation rate with time. The forms of the changes are linear in each time segment, which cannot reflect the real situation of the heat source passing through the plane. In addition, the value of V_{act} needs to be estimated first and then calibrated according to the experimental results. Continuous adjustment must be made until the simulated results are consistent with the experimental results.

An improved uniform body heat flux distribution is used to simulate the welding process. The action volume of the heat source is calculated according to the groove size and welding process parameters. As shown in Figure 7, assuming that all thermal energy is concentrated in the filler metal, V_{act} is determined by the product of the cross-sectional area of the welding layer (A) and the action length of the heat source (C). C is estimated by Equation (4) [31].

$$C = \tau \times U \times I \tag{4}$$

where τ is the coefficient determined by the welding method and welding process parameters, and in this paper $\tau = 3$ mm·kW^{-1}.

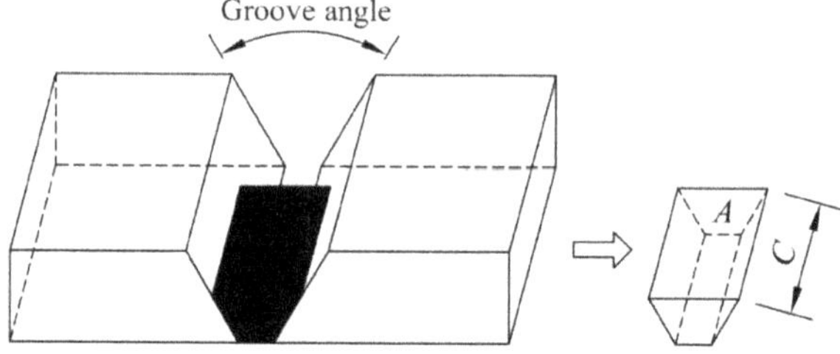

Figure 7. Action volume of the heat source.

Therefore, the total action time of the heat source is obtained by Equation (5).

$$t_{act} = \frac{C}{v} = \frac{\tau \times U \times I}{v} \tag{5}$$

where v is the welding speed.

Considering the effect of the heat source approaching and leaving the studied cross section during the welding process, the improved uniform body heat flux distribution changes with time instantaneously. The distributed heat flux is calculated by Equation (6) [32].

$$q = \frac{\eta \times U \times I}{V_{act}} \exp\left\{ \frac{-12 \times [v \times (t - t_0)]^2}{C^2} \right\} \tag{6}$$

where t is the transient loading time; t_0 is the time taken for the center of the heat source to move to the studied cross section, equal to half of t_{act}.

The modelling of the improved uniform body heat flux distribution is done by FORTRAN language written in DFLUX, one of user subroutines in ABAQUS. The arc efficiency factor is assumed to be 0.8 for the shielded metal arc welding (SMAW), and 0.6 for the gas tungsten arc welding (GTAW).

2.3. Boundary Conditions

The ambient temperature is assumed as 20 °C. The convection heat transfer exists between air and weldment surface during welding due to the temperature difference. Additionally, the temperature difference between weldment surface and surrounding environment will result in continuous radiation heat transfer. The convection heat transfer follows Newton's law of cooling (Equation (7)).

$$q_c = \alpha_c \times \Delta T \tag{7}$$

where q_c is the heat flux of convection heat transfer; α_c is the convection heat transfer coefficient; ΔT is the value of temperature difference between air and weldment surface.

The convection heat transfer coefficient is calculated by Equation (8), which is introduced into the FE model by the user subroutine FILM in ABAQUS 6.14-5.

$$\alpha_c = \begin{cases} 0.0668 \times T & 0 < T \le 500\ ^\circ\text{C} \\ 0.231T - 82.1 & T > 500\ ^\circ\text{C} \end{cases} \tag{8}$$

where T is the transient temperature of the weldment surface.

The heat flux radiated outward by weldment follows the Stefan–Boltzmann law (Equation (9)).

$$q = \varepsilon \times \sigma \times T^4 \tag{9}$$

where ε is the emissivity, assumed as 0.85; σ is the Stefan-Boltzmann constant with the value of 5.67×10^{-8} W/m^2·K^4. T is the temperature of the weldment surface.

The radiation heat transfer between weldment surface and surrounding environment is calculated by Equations (10) and (11).

$$q_r = \alpha_r \times \left(T - T_f \right) \tag{10}$$

$$\alpha_r = \varepsilon \times \sigma \times \frac{\left(\frac{T+273}{100} \right)^3 - \left(\frac{T+273}{100} \right)^4}{T - T_f} \tag{11}$$

where q_r is the heat flux of radiation heat transfer; α_r is the radiation heat transfer coefficient; T_f is the ambient temperature.

During the mechanical analysis, degrees of freedom for node A and node B at two ends of the bottom surface of the FE model (shown in Figure 5) are constrained in the X direction and Y direction to prevent the rigid body displacement.

2.4. Thermal Loading Patterns

Welding linear energy is the heat energy input by welding heat source to unit length welding seam, which influences the surface forming of welding seam and the formation of welding defects. The magnitude of welding linear energy is calculated by Equation (12).

$$Q = \frac{\eta \times U \times I}{v} \tag{12}$$

Three different repair welding linear energy levels will be considered in the fracture analysis. The values of three selected linear energies are 10 kJ/cm, 16 kJ/cm and 25 kJ/cm, respectively. According to the heat flux distribution defined by Equation (6), the time distribution curves of heat generation rate in the weld repair area under three different linear energy rates is plotted in Figure 8. The heat generation rate presents a Gaussian profile with time, which can reflect the moving features of the heat source. With the increase of linear energy, both the duration of the thermal loading and the generated total heat energy increase. The duration of the thermal loading is 3.96 s for the linear energy of 10 kJ/cm, 6.34 s for the linear energy of 16 kJ/cm and 9.90 s for the linear energy of 25 kJ/cm. All three thermal loading patterns have the same peak heat generation rate in the intermediate instant of the corresponding loading time, which is 14.8 GW/m^3. The intermediate instant is the time when the center of the heat source passes through the studied cross section.

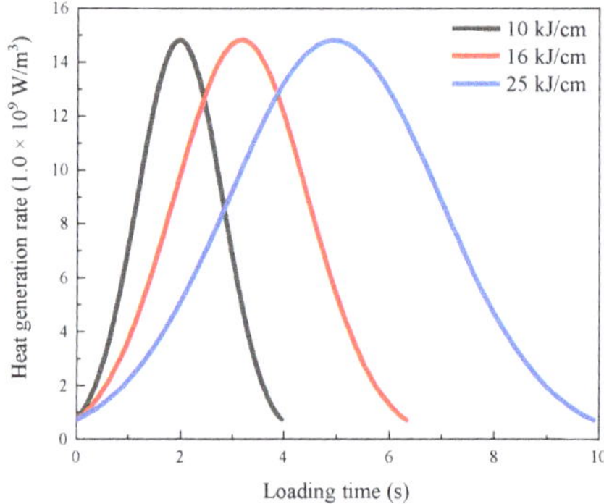

Figure 8. Time distribution of heat generation rate under different linear energy.

2.5. Damage Model

The XFEM used in this paper is based on the cohesive crack model [33]. The damage response of a crack takes the traction–separation law as the constitutive relation. As shown in Figure 9, the damage development includes three stages: damage initiation, damage evolution and element failure.

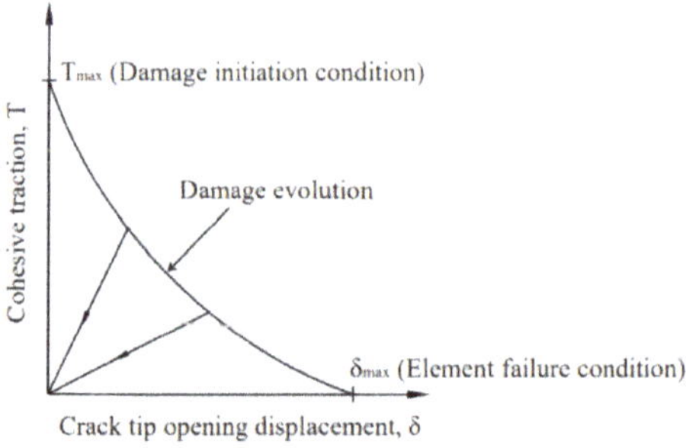

Figure 9. The traction-separation law.

The damage initiation of a crack is the starting point at which the cohesive stiffness between the crack surfaces begins to degrade. It occurs when the stress or strain of the element reaches a critical value. Here, the maximum principal stress (MPS) criterion is used to judge whether a crack reaches the damage initiation condition. Meanwhile, the crack propagation direction is also controlled by the MPS criterion. The cohesive crack will propagate along the direction of maximum principal stress. The damage initiation criterion satisfies Equation (13).

$$1 \leq f = \frac{\langle \sigma_{max} \rangle}{\sigma_{max}^0} \leq 1 + f_{tol} \tag{13}$$

where $< \sigma_{max} >$ is the actual maximum principal stress of the element, the symbol "$< >$" means the damage initiation does not exist in a pure compression state; σ_{max}^0 is the allowable maximum principal stress defined; f_{tol} is the tolerance with a value of 0.05, which is a default value recommended by ABAQUS.

The damage evolution defines the degradation patterns of cohesive stiffness after damage initiation. The nonlinear exponential softening response is adopted to describe the degradation of cohesive stiffness for the analysis of elastoplastic fracture mechanics (EPFM). As the fracture energy is one of the fracture properties of materials, the evolution law based on energy method is selected. Once the crack propagation driving force exceeds the equivalent critical energy release rate (equivalent to fracture energy), the crack will extend. The equivalent critical energy release rate is calculated by the Benzeggagh–Kenane law (Equation (14)) [34].

$$G_{eqC} = (G_{IIC} - G_{IC}) \times \left(\frac{G_{II} + G_{III}}{G_I + G_{II} + G_{III}} \right)^\eta + G_{IC} \tag{14}$$

where G_{eqC} is the equivalent critical energy release rate; G_I, G_{II} and G_{III} are the energy release rates of mode I, II and III cracks, respectively; G_{IC} and G_{IIC} are the critical energy release rates of mode I and II cracks, respectively; η is the exponent, to which the response is insensitive for isotropic failure.

Fracture toughness is an index to measure the capacity of materials to prevent crack propagation, which is one of the inherent properties of materials. It can be expressed by a single parameter such as energy release rate G, stress intensity factor K, crack tip opening displacement CTOD and J-integral. The critical energy release rate is used as the fracture toughness here. The fracture failure mode is assumed to be isotropic, so $G_{IC} = G_{IIC}$. The fracture energy in the process of damage evolution is calculated on the basis of the fracture toughness of P91 steel. The tensile strength of P91 steel is taken as an approximation for the MPS. For both fracture toughness and tensile strength, temperature-dependent data are used to simulate the actual fracture behavior. The fracture parameters under high temperature are extrapolated by a linear interpolation method. The fracture properties of P91 steel are shown in Figure 10.

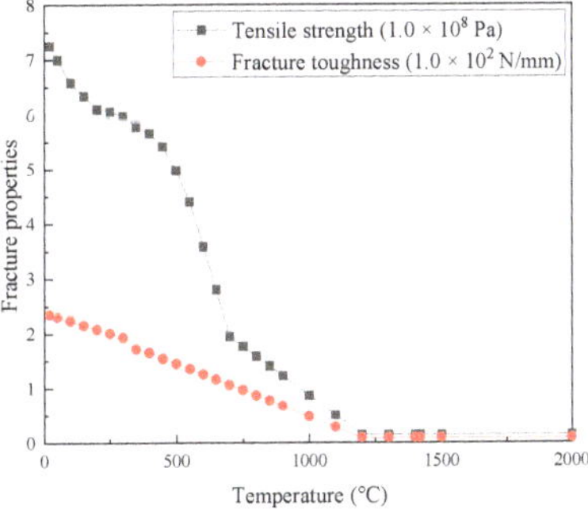

Figure 10. Temperature-dependent fracture properties for P91 steel, used in the FE simulation, data from [35,36].

With the development of damage evolution, the cohesive traction between crack surfaces decreases. When the cohesive traction is reduced to zero, the element fails, which means that the crack surfaces have been completely opened.

The fracture problems with damage definition have strong nonlinearity and discontinuity, which makes numerical solutions difficult to converge. ABAQUS provides a viscous regularization method to solve this problem. By setting an appropriate small viscosity coefficient, the convergence of the fracture model will be significantly improved. Here, the viscosity coefficient is set to be 1.0×10^{-5}.

The type IV crack with a size of 0.5 mm perpendicular to the surface is prefabricated in the HAZ of repair welding, as shown in Figure 11a. *d* is the distance between the crack and the fusion line. The initial crack is introduced into the FE model by the XFEM technique, and the contact attributes between the crack surfaces are set as frictionless for tangential behavior and "hard" contact for normal behavior, as shown in Figure 11b.

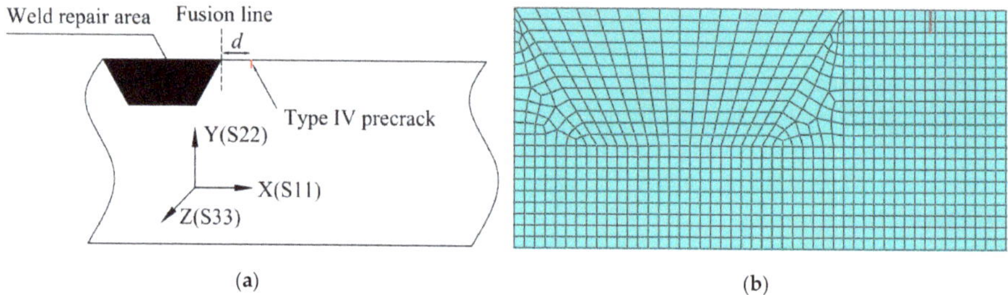

Figure 11. Precrack definition: (**a**) the physical model with a precrack; (**b**) the XFEM model with a precrack.

3. Weld Repair Experiments

3.1. Weld Repair Specimens

Weld repair experiments of the P91 steel welded joint were carried out to verify the feasibility of the FE model. The size of the specimens was the same as that of the FE model. The chemical composition of experimental base metal is shown in Table 1, which measures up to ASME BPVC.II.A-2019 [37]. The selected weld metal was particularly suited for matching P91 steel. ER90S-B9 welding rod was used as a filler metal for GTAW, while E9015-B9 stick electrode was used as a filler metal for SMAW. The chemical composition of the weld metal is shown in Table 2, which measures up to ASME BPVC.II.C-2019 [38].

Table 1. Chemical composition of P91 steel (wt. %).

C	Si	Mn	S	P	Cr	Ni	Mo	V	Nb	N	Al
0.08	0.27	0.6	0.006	0.007	8.86	0.38	0.98	0.19	0.06	0.06	0.04

Table 2. Chemical composition of the weld metal (wt. %).

Material Grade	C	Si	Mn	Cr	Ni	Mo	V	Nb
ER90S-B9	0.1	0.3	0.5	9.0	0.7	1.0	0.2	0.06
E9015-B9	0.09	0.2	0.6	9.0	0.8	1.1	0.2	0.05

Firstly, the base metal plates were connected by multilayer welding. GTAW was used for root welding, followed by SMAW used for filler welding and cover welding. The shielding gas for GTAW was argon. Then the welded joint was air-cooled to the ambient temperature. The metal near the FGHAZ of the welded joint, where it was prone to IV type cracking, was removed. Finally, repair welding technique was utilized to refill it by SMAW with the welding linear energy of 10 kJ/cm. The specific welding parameters for each layer and weld repair are shown in Table 3. The interlayer temperature was maintained at 200–300 °C during welding. The morphology of welding seams after initial welding and repair welding is shown in Figure 12.

Table 3. Specific welding parameters for each layer and weld repair.

Layer Number	Welding Method	Electrode Grade	Electrode Diameter (mm)	Welding Current (A)	Arc Voltage (V)	Welding Speed (cm/min)
1	GTAW	ER90S-B9	2.4	100	12	6
2	GTAW	ER90S-B9	2.4	110	14	6
3	SMAW	E9015-B9	3.2	110	20	12
4	SMAW	E9015-B9	3.2	120	22	12
Repair	SMAW	E9015-B9	3.2	120	22	12

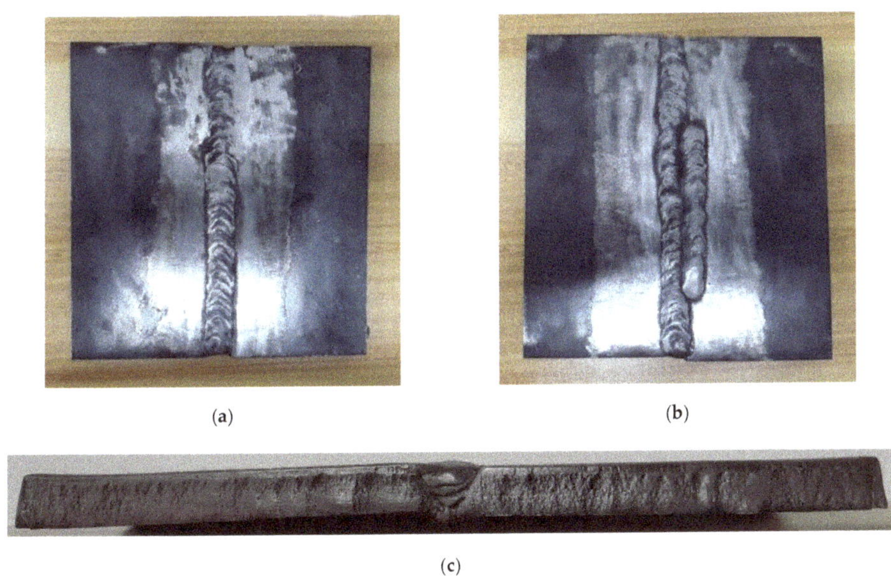

(a)

(b)

(c)

Figure 12. Morphology of specimens: (**a**) after initial welding; (**b**) after repair welding; (**c**) morphology of welding layers.

3.2. Verification of the FE Model

After the fabrication of weld repair specimens, the impact indentation strain method [39] was used to measure the residual stress. A spherical indenter was used to generate indentations in the form of impact loads at the location of measuring points. On the one hand, the impact action makes the material appear to have plastic flow deformation. On the other hand, the elastoplastic deformation caused by indentations itself changes under the action of residual stress. The total deformation after superposition of the two kinds of deformation is called strain increment. The relationship between the strain increment generated by the impact indentation strain method and elastic strain satisfies Equation (15).

$$\Delta\varepsilon = a_1 \times \varepsilon_e + a_2 \times \varepsilon_e{}^2 + a_3 \times \varepsilon_e{}^3 + B \tag{15}$$

where $\Delta\varepsilon$ is the strain increment; ε_e is the elastic strain; a_1, a_2, a_3 are the coefficients obtained from the calibration curve; B is the strain increment in a zero-stress state.

The biaxial resistance strain gauge containing two sensing grid elements with different axial directions, namely X-axis direction and Z-axis direction, was used to measure residual strain. A certain size of indentation was produced at the midpoint of the grid axis by the impact load. The value of strain increment was recorded by the strain recording instrument. The value of residual strain was determined according to the relationship between the calibrated elastic strain and the strain increment. The value of residual stress was calculated by Hooke's law.

The surface residual stress of specimens after both initial welding and repair welding was measured. Five measuring points were taken from each specimen, and the distance between adjacent measuring points on each side of the welding seam was 2.5 mm. The specific position of each measuring point is shown in Figure 13. The measurement of surface residual stress was finished by the KJS-3\4 indentation stress measurement system (Developed by Institute of Metal Research, Chinese Academy of Sciences, Shenyang, China). Before measurements, the accuracy of the indentation stress measurement system was calibrated with conventional residual stress measurement methods, such as X-ray diffraction, which measures up to GB/T 24179-2009. As shown in Figure 14a, the measurement system consists of two parts: portable intelligent stress tester and indentation generating device. The indentation generating device includes a striking rod (for generating indentation), a permanent magnet fixed base (for restricting the displacement of the measurement device) and a centering microscope (for centering the grid axis of strain gauge). The residual stress on-site measuring is shown in Figure 14b.

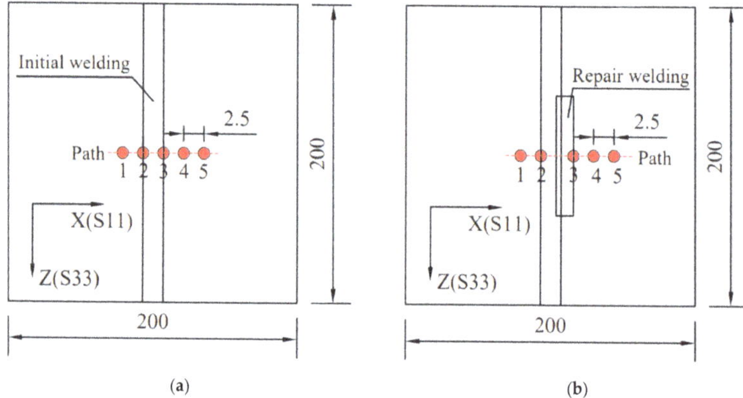

(a) (b)

Figure 13. Position of measuring points for specimens (unit: mm): (**a**) after initial welding; (**b**) after repair welding.

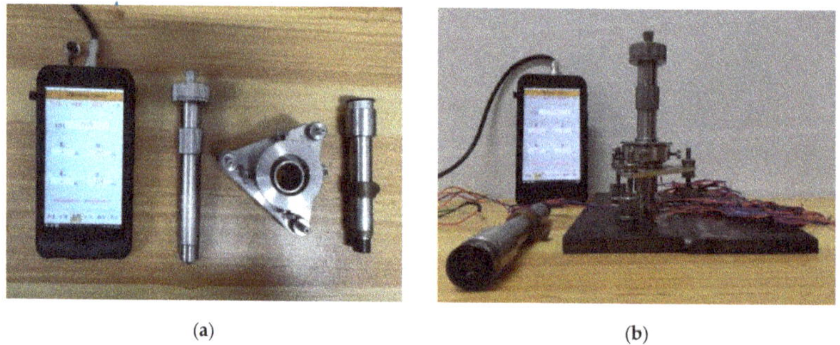

(a) (b)

Figure 14. The measurement of residual stress: (**a**) residual stress measurement system; (**b**) residual stress on-site measuring.

Figure 15 presents a comparison of the residual stress within and around the welding seam zone obtained from experiments and FE simulations. It is shown that the simulated results are in good agreement with the experimental results. The maximum difference between the simulated value and the experimental value is less than 10%. Therefore, the FE model and program developed in this paper are suitable for the thermal-mechanical coupled simulation of repair welding process for the welded joint. The reason for the existing difference between the simulated results and the experimental results is that the 2-D numerical model is a simplification of the 3-D experimental model. The 2-D plane in the FE simulations actually corresponds to the middle plane of an infinite plate. Thus, only when the size of specimens in the length direction is infinitely large, the simulated results may be very close to the experimental results.

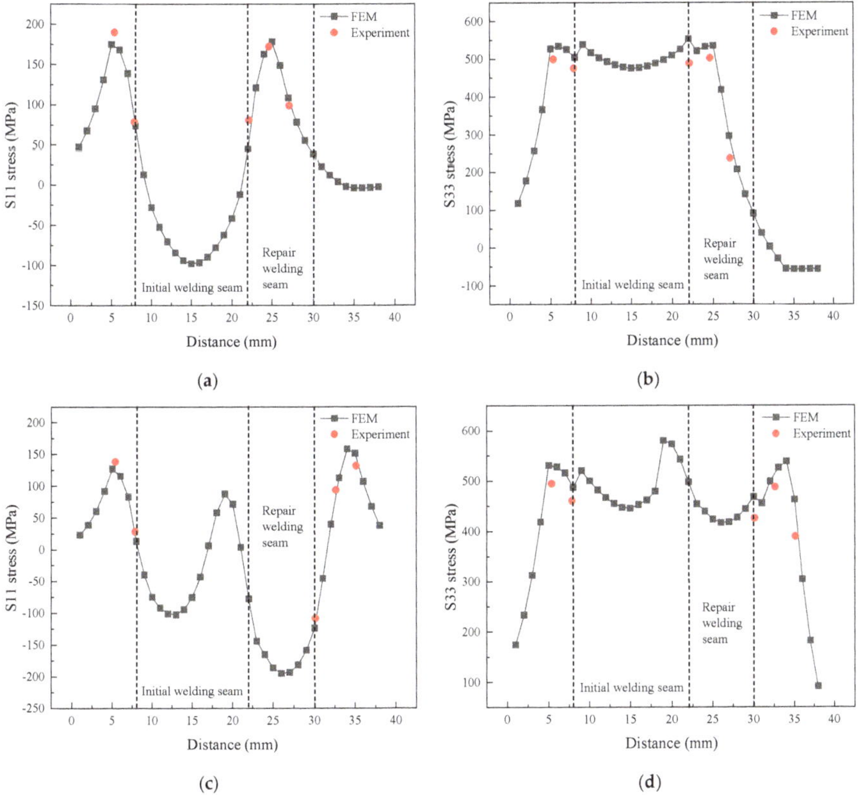

Figure 15. Comparison of the residual stress obtained from FE simulations and experiments: (**a**) S11 stress after initial welding; (**b**) S33 stress after initial welding; (**c**) S11 stress after repair welding; (**d**) S33 stress after repair welding.

4. Results and Discussion

4.1. Thermal Analysis

The transient temperature contours at the intermediate instant and the end instant of repair welding for different linear energy are shown in Figure 16. The variation of linear energy has a significant influence on the temperature profile of the welding seam and HAZ. With the increase of linear energy, the peak temperature at the intermediate instant of repair welding increases. The peak temperature is 1819 °C for the linear energy of 10 kJ/cm, 2088 °C for the linear energy of 16 kJ/cm and 2320 °C for the linear energy of 25 kJ/cm. This is because higher linear energy involves longer loading times, thus more thermal energy is poured into the bulk metal during repair welding. After the intermediate instant of repair welding, due to the fact that the heat source starts to depart from the cross section, the heat generation rate decreases continuously. At the end instant of repair welding, the peak temperature for different linear energy has no obvious difference, all are around 1500 °C. In addition, with the increase of repair welding linear energy, the range of the HAZ expands. The higher linear energy brings higher overall temperature of the HAZ and poorer fracture properties.

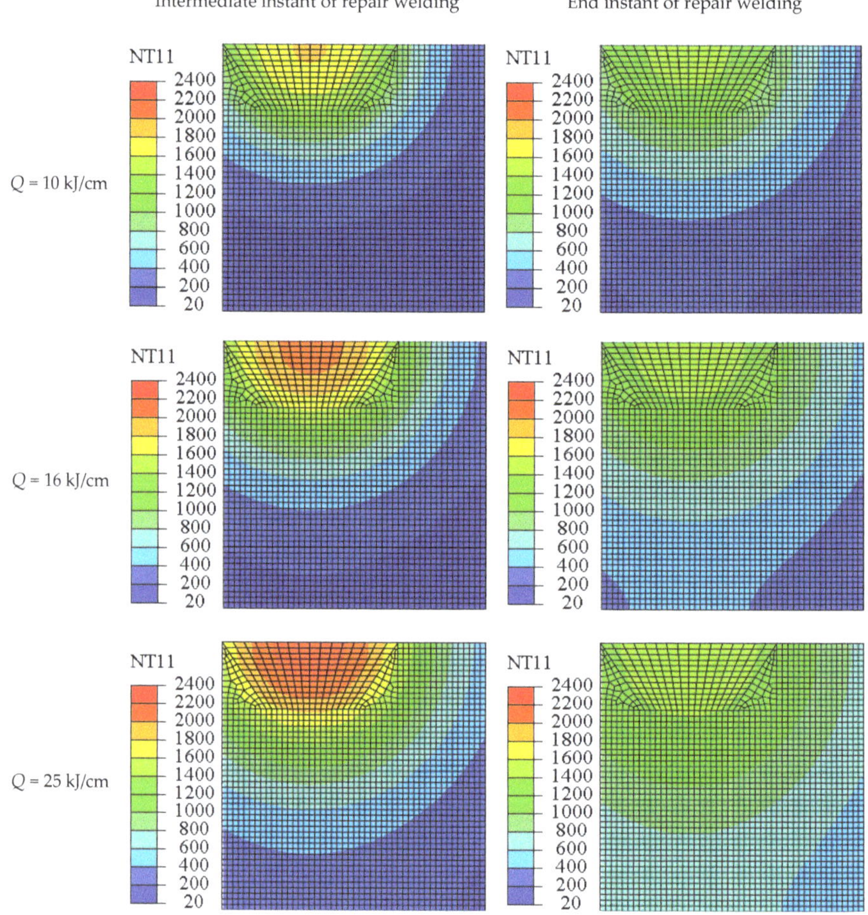

Figure 16. Temperature contours at the intermediate instant and the end instant of repair welding for different linear energy.

4.2. Mechanical Analysis

The calculated temperature field loads are used as the predefined field of stress calculation to perform the mechanical analysis. In Figure 18, the contours of the Von Mises stress field are shown at the end instant of repair welding and the end instant of cooling (the transient time at which the weld repaired component is cooled to the ambient temperature) for different linear energy. With the increase of linear energy, the range of high stress area in the HAZ at both the end instant of repair welding and the end instant of cooling expands. For different linear energy, the peak Von Mises stress in the HAZ at the end instant of repair welding and the end instant of cooling is similar. The peak Von Mises stress at the end instant of repair welding is about 445 MPa, and the peak Von Mises stress at the end instant of cooling is about 480 MPa. At the end instant of cooling, the peak Von Mises stress of the repair welding seam is greater than that of the HAZ. This is because of the different material properties considered between the weld metal and the base metal.

Figure 17 shows the temperature distribution at a depth of 0.5 mm in the HAZ at the end instant of repair welding for different linear energy. The overall temperature in the HAZ increases with the increment of linear energy. The peak temperature of the HAZ is 961 °C for the linear energy of 10 kJ/cm, 1130 °C for the linear energy of 16 kJ/cm and 1258 °C for the linear energy of 25 kJ/cm. The temperature distribution of the HAZ is gentler for high linear energy.

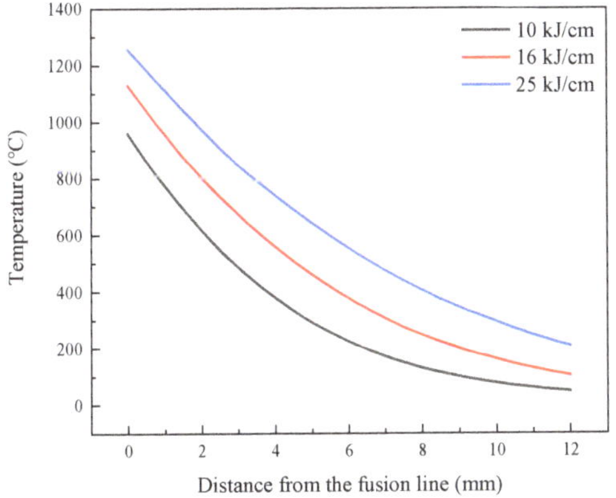

Figure 17. Temperature distribution at a depth of 0.5 mm in the heat affected zone (HAZ) at the end instant of repair welding for different linear energy.

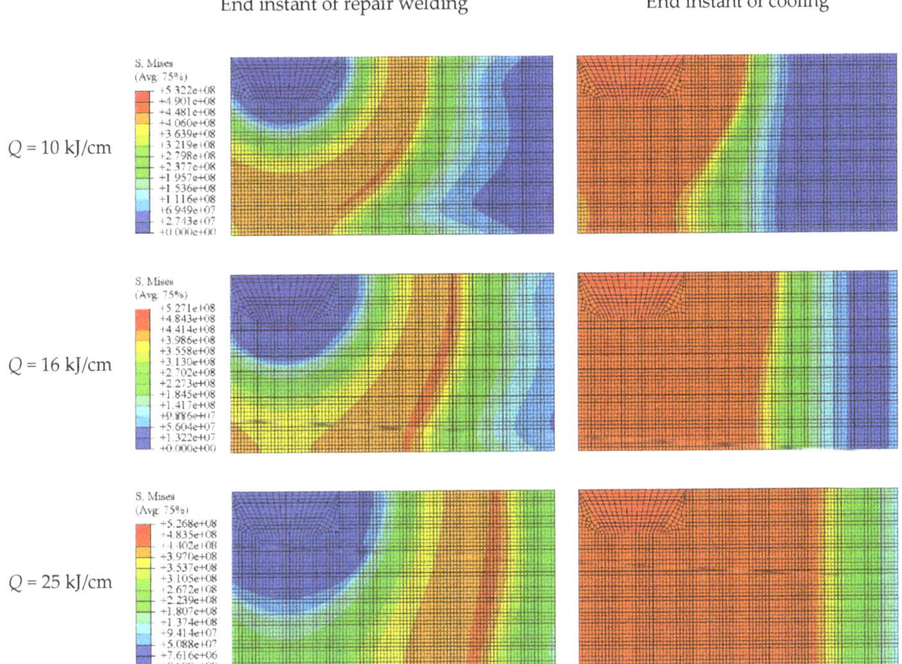

Figure 18. Stress contours at the end instant of repair welding and the end instant of cooling for different linear energy.

The S11, S33 and Von Mises stresses at a depth of 0.5 mm in the HAZ at the end instant of repair welding for different linear energy are shown in Figure 19. With the increase of linear energy, the range of high stress area of S11, S33 and Von Mises expands and moves towards the direction far away from the fusion line. The peak S33 stress and Von Mises stress has little change with the increase of linear energy, while the peak S11 stress increases slightly with the increment of linear energy. The S11 stress and S33 stress are tensile stress near the fusion line and compressive stress far away from the fusion line.

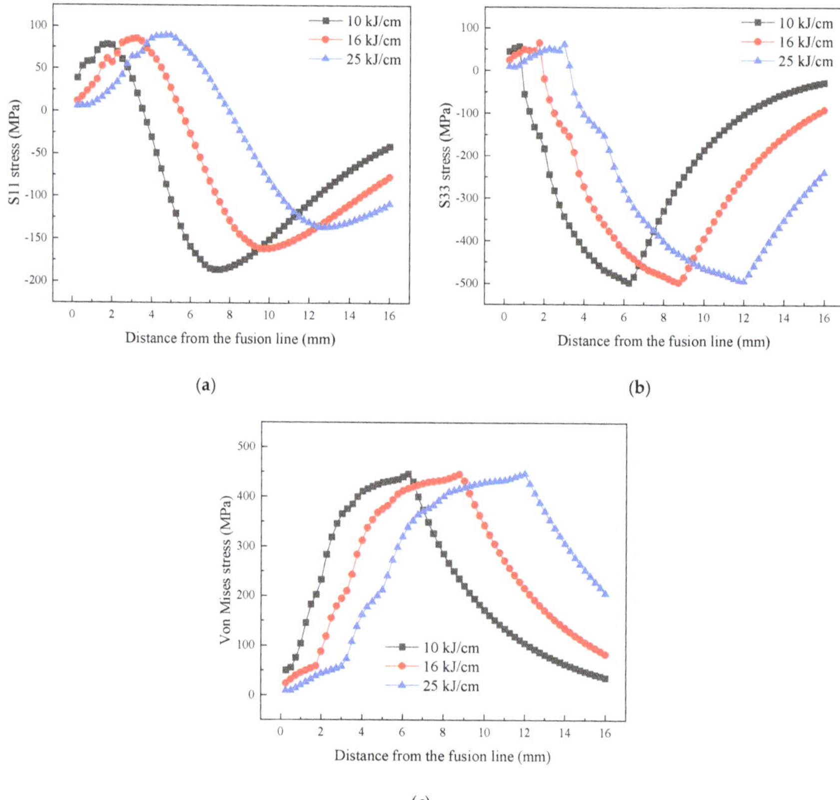

Figure 19. Stress distribution at a depth of 0.5 mm in the HAZ at the end instant of repair welding for different linear energy: (**a**) S11 stress in the HAZ; (**b**) S33 stress in the HAZ; (**c**) Von Mises stress in the HAZ.

Figure 20 shows the S11, S33 and Von Mises stress distribution at a depth of 0.5 mm in the HAZ at the end instant of cooling for different linear energy. Both the S33 stress and Von Mises stress have similar peaks for different linear energy. The peak S33 stress is about 525 MPa, and the peak Von Mises stress is about 480 MPa. In the cooling process, the S11 stress near the fusion line in the HAZ changes from tensile stress to compressive stress, and the S11 stress far away from the fusion line changes from compressive stress to tensile stress. Almost all the S33 stress in the HAZ changes from compressive stress to tensile stress, and the peaks change from −500 MPa to +525 MPa, which is caused by the shrinkage of welding seam during cooling. The remarkable tensile stress may lead to the propagation of initial cracks in the HAZ. Moreover, with the increase of linear energy, the range of the high stress area in the HAZ expands. The range of the high stress area in the HAZ is 4 mm for the linear energy of 10 kJ/cm, 6 mm for the linear energy of 16 kJ/cm and 10 mm for the linear energy of 25 kJ/cm.

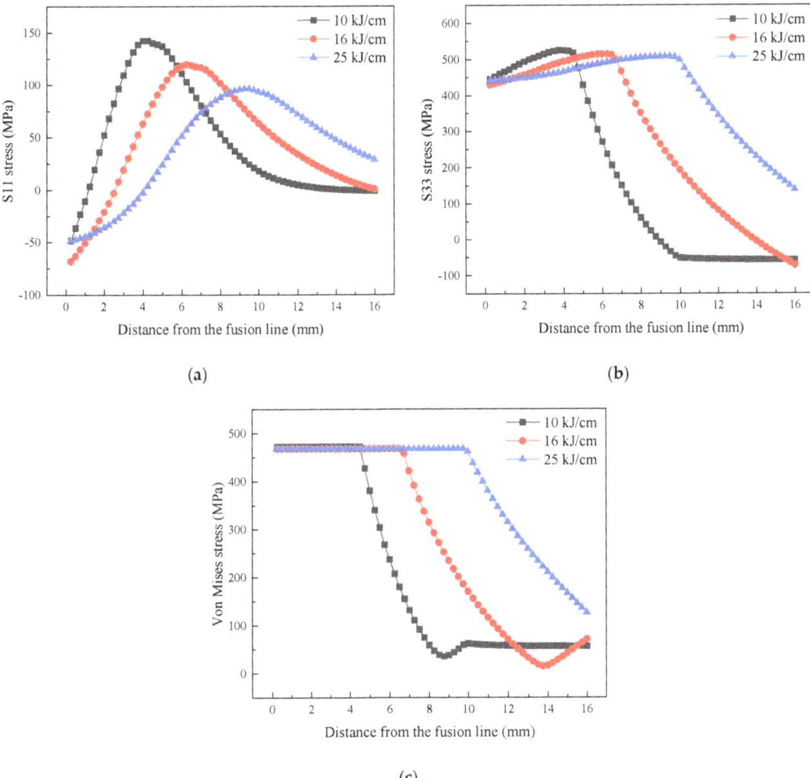

Figure 20. Stress distribution at a depth of 0.5 mm in the HAZ at the end instant of cooling for different linear energy: (**a**) S11 stress in the HAZ; (**b**) S33 stress in the HAZ; (**c**) Von Mises stress in the HAZ.

4.3. XFEM Analysis

Under repair welding thermal shock, the welding seam and HAZ deform plastically, so the fracture mode is elastoplastic fracture. The whole fracture process has strong nonlinearity, including not only the nonlinearity of material properties, but also the geometric nonlinearity consisting of large thermal deformation and crack propagation.

The crack at the position of $d = 2$ mm under the repair welding linear energy of 10 kJ/cm is selected to analyze and the simulated crack propagation paths are shown in Figure 21. It is noteworthy that the crack propagation occurs during the cooling process, which is in agreement with the mechanical analysis. The crack propagation direction is almost perpendicular to the surface due to the control of MPS criterion and there is a certain tendency for the crack to deflect to the welding seam during the propagation process. A total of two elements are extended and the extracted crack propagation length is 0.52 mm. Figure 21a shows the MPS contours for the crack zone at different transient times. $t = 3.96$ s is the end instant of repair welding and the start instant of cooling. At the start instant of cooling, the MPS around the crack tip is lower than the allowable MPS at the corresponding temperature according to the damage initiation criterion, so the crack does not propagate. With the cooling process going on, the tensile stress increases continuously, which leads to the increment of the MPS around the crack tip. At $t = 4.43$ s, the crack damage initiation condition is reached, and the new crack surface begins to form. After the formation of the new crack surface, the stress concentration around the crack tip is released and the MPS decreases. A new driving force is required to promote the further crack propagation. At $t = 5.22$ s, the new driving force makes the crack extend further.

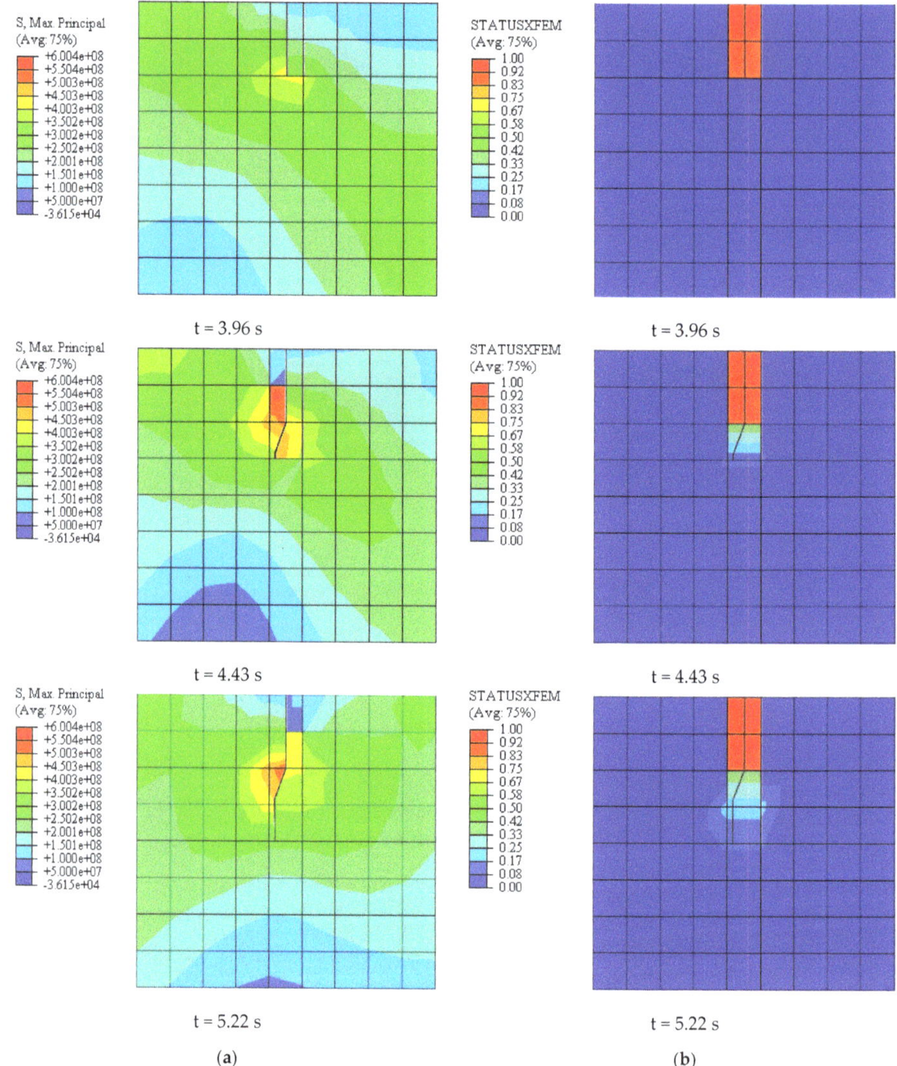

Figure 21. Crack propagation paths at different transient times (d = 2 mm, Q = 10 kJ/cm): (**a**) maximum principal stress (MPS) contours; (**b**) status of the enriched element (STATUSXFEM).

Figure 21b shows the status of the enriched element (STATUSXFEM) for the crack zone at different transient times. The variable STATUSXFEM represents the damage degree of the crack, ranging from 0 to 1.0. A value of 0 means that the material has no damage. A value of 1.0 means that the crack is fully opened and the cohesive traction between the crack surfaces is zero. When the value is greater than 0 and less than 1.0, it means that the cohesive crack has already extended and needs extra driving force to fully open. Although the new cracks are not fully opened, in the complex service environment, the cohesive cracks containing damage are likely to open completely after being driven by the extra driving force, which threatens the structural integrity of weld repaired components.

According to the crack propagation paths in Figure 21, two elements in the initial crack front are selected to analyze. The transient S11, S33 and Von Mises stresses of the elements under repair welding thermal shock are shown in Figure 22. The first 3.96 s of transient time is the repair welding process. In this process, the HAZ is compressed due to the outward expansion of the welding seam. The S11 and S33 stresses are compressive stress, and the value of compressive stress increases with the promoting of repair welding. After repair welding, the welding seam begins to cool and shrink, which results in the tensile stress in the HAZ. The S11 and S33 stress changes from compressive stress to tensile stress in the early stage of cooling and the crack propagation occurs in this stage.

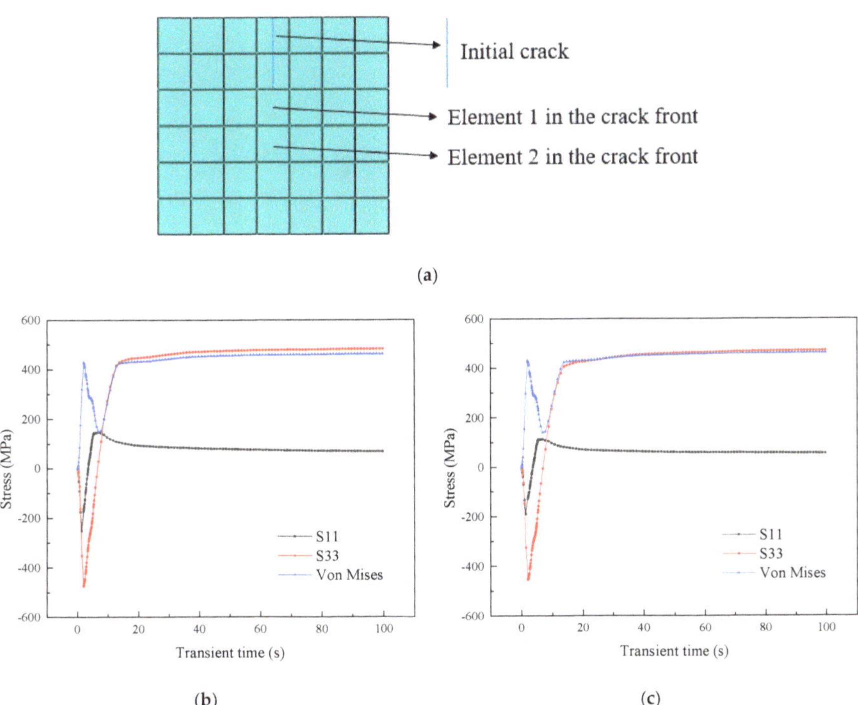

Figure 22. Transient stress curves of the elements in the crack front: (**a**) initial crack and elements in the crack front; (**b**) element 1; (**c**) element 2.

In order to investigate the influence of repair welding thermal shock on the cracks at different distances from the fusion line for three different linear energy values, the initial cracks with a size of 0.5 mm perpendicular to the surface were prefabricated at the positions where the distances from the fusion line were 2 mm, 4 mm, 6 mm, 8 mm and 10 mm, respectively. Figure 23 shows the crack propagation predicted by XFEM at different positions for three different linear energy values. All the crack propagation occurs in the early stage of cooling. For cracks at the same location, with the increase of linear energy, the crack propagation length and the number of damaged elements increase. This is because the higher linear energy brings the higher temperature of the HAZ, which leads to poorer crack resistance. The longest crack propagation appears at the position of $d = 2$ mm under the linear energy of 25 kJ/cm. The extracted crack propagation length is 1.1 mm, which is almost twice of the initial crack length. In this case, the fully opened crack propagation has occurred, and the crack is likely to develop into the macrocrack, even penetrate through the whole wall thickness. For the same linear energy, the crack propagation length decreases with the increment of distance from the fusion line. Both the cracks with $d > 6$ mm for the linear energy of 10 kJ/cm and the cracks with $d > 8$ mm for the linear energy of 16 kJ/cm do not propagate. The reason is that the lower the linear energy is, the smaller the scope of the HAZ is. Once the crack is far away from the fusion line, the driving force is not enough to promote the crack propagation. It should be noted that the area where cracks propagate is the high stress area where the peak stress is located. Almost all the directions of crack propagation are perpendicular to the surface and tend to deflect to the welding seam, which may be related to the high stress in the welding seam.

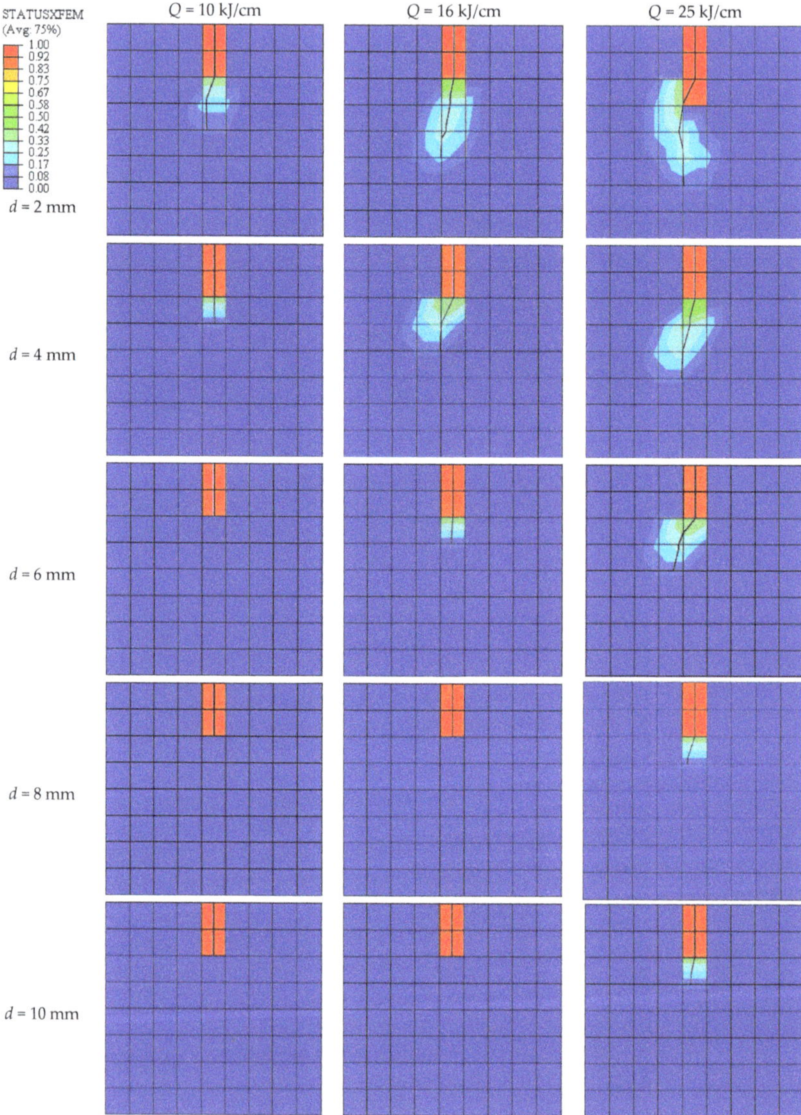

Figure 23. Crack propagation predicted by XFEM at different positions for three different linear energy values.

Since the crack resistance of materials is related to temperature, the temperature at the crack tip is one of the key factors affecting crack propagation. On the one hand, the higher the temperature is, the poorer the crack resistance is, and the cracks are easier to extend under the same driving force. On the other hand, the larger the temperature gradient is, the greater the thermal stress is, so the driving force is sufficient for crack propagation. Figure 24 shows the thermal cycle curves at different crack tips in the HAZ for three different linear energy values. For the same linear energy, with the decrease of the distance from the fusion line, the difference of the peak temperature at the crack tip becomes larger. As a result, the closer the crack is to the fusion line, the greater the temperature gradient and temperature change around the crack tip, the greater the driving force and the longer the crack propagation length. For the same crack tip, the peak temperature increases with the increment of linear energy. Therefore, the greater the linear energy is, the poorer the crack resistance of materials in the HAZ is. In the early stage of cooling, the temperature value in the HAZ is higher and the temperature difference in the HAZ is larger, so crack propagation occurs in the early stage of cooling. With the cooling process going on, the temperature at the crack tip at different positions gradually tends to become consistent.

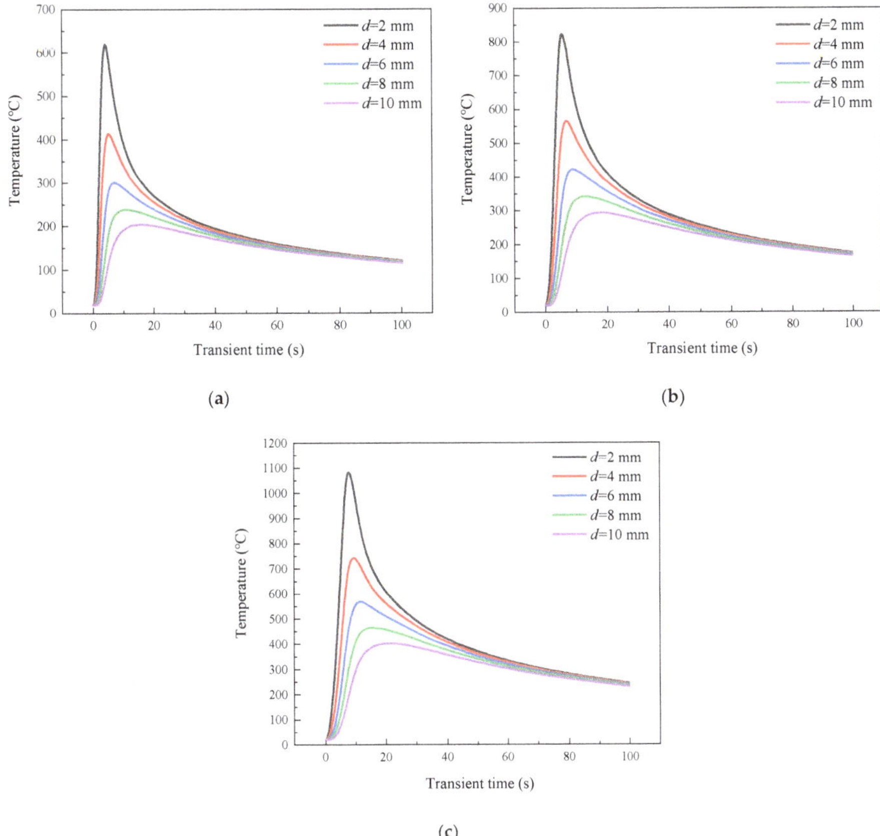

Figure 24. Thermal cycle curves at crack tips in the HAZ for three different linear energy values: (**a**) $Q = 10$ kJ/cm; (**b**) $Q = 16$ kJ/cm; (**c**) $Q = 25$ kJ/cm.

The distribution of heat flux through the crack surfaces with the transient time is shown in Figure 25. The distribution of heat flux is similar to the thermal cycle curves at crack tips, which indicates that the temperature at crack tips is proportional to the thermal energy density through the crack surfaces. For the same crack surface, the peak heat flux increases with the increment of linear energy. For the same linear energy, in repair welding process and early stage of cooling, the closer the crack surface is to the fusion line, the greater the heat flux gradient is and the greater the heat flux change is.

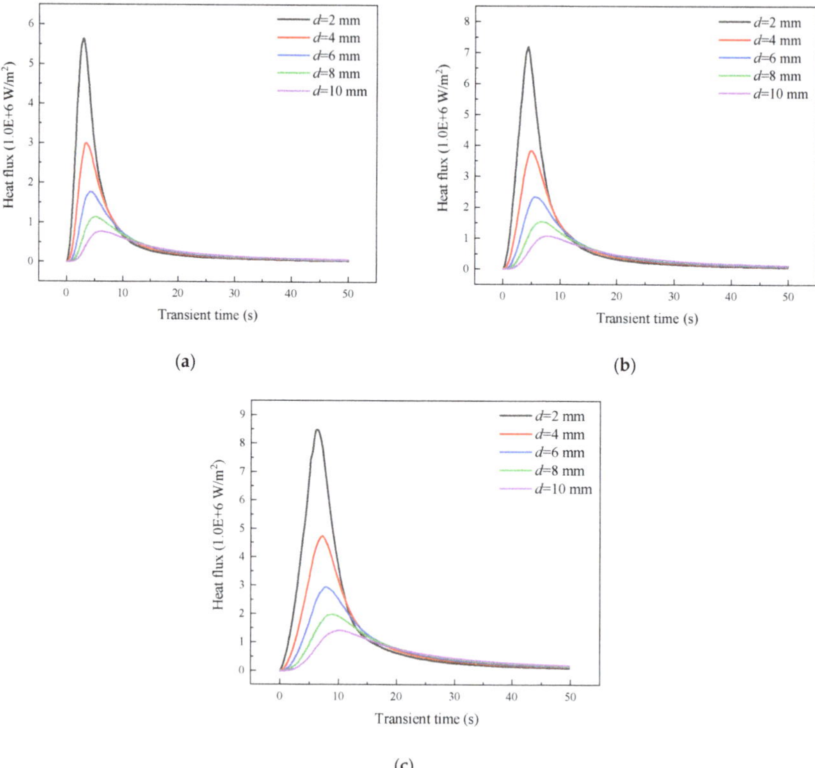

Figure 25. Heat flux distribution through crack surfaces in the HAZ for three different linear energy values: (**a**) Q = 10 kJ/cm; (**b**) Q = 16 kJ/cm; (**c**) Q = 25 kJ/cm.

5. Conclusions

In this study, the XFEM based on a cohesive crack model is used to establish the repair welding model of P91 steel welded joint containing initial cracks. The MPS criterion and energy method are used to define the damage initiation and damage evolution, respectively. Considering the temperature-dependent material physical properties and fracture properties, the effect of repair welding thermal shock on the crack propagation behavior in the HAZ is simulated. The influence of different repair welding linear energy and different crack positions on the cracking features is analyzed. The main conclusions are as follows:

(1) With the increase of linear energy, the range of the HAZ expands and the overall temperature of the HAZ rises. With the increase of linear energy, the range of S11, S33 and Von Mises high stress area expands.

(2) In the repair welding and cooling process, the material in the HAZ undergoes a change from compressed state to tensioned state due to the expansion and shrinkage of the welding seam. It is the direct cause of crack propagation in the HAZ.

(3) Crack propagation occurs in the early stage of cooling. The crack propagation direction is roughly perpendicular to the upper surface and has a tendency to deflect to the welding seam.

(4) For cracks at the same position, with the increase of linear energy, the crack propagation length increases. For the same linear energy, with the distance from the fusion line increases, the crack propagation length decreases.

(5)   After a certain distance from the fusion line, the cracks in the HAZ do not extend. This distance is the critical distance for the crack propagation. With the increase of linear energy, the critical distance of crack propagation increases. The critical distance of crack propagation is consistent with the high stress area after repair welding. Therefore, for repair welding, low linear energy is preferred to constrain the cracking behavior in the HAZ.

Author Contributions: Conceptualization, J.Z.; methodology, K.Y.; software, K.Y.; validation, Y.Z.; formal analysis, K.Y.; investigation, Y.Z.; resources, Y.Z.; data curation, K.Y.; writing—original draft, K.Y.; writing—review & editing, K.Y., Y.Z. and J.Z.; visualization, K.Y.; supervision, J.Z.; project administration, K.Y.; funding acquisition, J.Z. All authors have read and agreed to the published version of the manuscript.

Funding: This research was funded by the National Key Research and Development Program of China (2016YFC0801905).

Acknowledgments: The authors gratefully acknowledge the financial support from the National Key Research and Development Program of China (2016YFC0801905). Furthermore, the authors are grateful for the technical support of Jiangsu Key Lab of Design and Manufacture of Extreme Pressure Equipment.

Conflicts of Interest: The authors declare no conflict of interest.

References

1. Viswanathan, R.; Coleman, K.; Rao, U. Materials for ultra-supercritical coal-fired power plant boilers. *Int. J. Pres. Vessel Pip.* **2006**, *83*, 778–783. [CrossRef]
2. Klueh, R.L. Elevated temperature ferritic and martensitic steels and their application to future nuclear reactors. *Int. Mater. Rev.* **2005**, *50*, 287–310. [CrossRef]
3. Chu, Q.L.; Zhang, M.; Li, J.H.; Chen, Y.N.; Luo, H.L.; Wang, Q. Failure analysis of a steam pipe weld used in power generation plant. *Eng. Fail. Anal.* **2014**, *44*, 363–370. [CrossRef]
4. Hu, Z.X.; Zhao, J.P. Effects of martensitic transformation on residual stress of P91 welded joint. *Mater. Res. Express* **2018**, *5*, 096528. [CrossRef]
5. Abson, D.J.; Rothwell, J.S. Review of type IV cracking of weldments in 9–12%Cr creep strength enhanced ferritic steels. *Int. Mater. Rev.* **2013**, *58*, 437–473. [CrossRef]
6. Viswanathan, R.; Gandy, D. Weld Repair of Aged Cr-Mo Steel Piping—A Review of Literature. *J. Press. Vessel Technol. ASME* **2000**, *122*, 76–85. [CrossRef]
7. Parker, J.D.; Siefert, J.A. Weld repair of Grade 91 piping and components in power generation applications, creep performance of repair welds. *Mater. High. Temp.* **2016**, *33*, 58–67. [CrossRef]
8. Weglowski, M. Determination of GTA and GMA welding arc temperatures. *Weld. Int.* **2005**, *19*, 186–192. [CrossRef]
9. Alvarez, P.; Vázquez, L.; Ruiz, N.; Rodríguez, P.; Magaña, A.; Niklas, A.; Santos, F. Comparison of hot cracking susceptibility of TIG and laser beam welded alloy 718 by varestraint testing. *Metals* **2019**, *9*, 985. [CrossRef]
10. Chelladurai, A.M.; Gopal, K.A.; Murugan, S.; Albert, S.K.; Venugopal, S.; Jayakumar, T. Effect of energy transfer modes on solidification cracking in pulsed laser welding. *Sci. Technol. Weld. Join.* **2015**, *20*, 578–584. [CrossRef]
11. Hosseini, S.A.; Abdollah-zadeh, A.; Naffakh-Moosavy, H.; Mehri, A. Elimination of hot cracking in the electron beam welding of AA2024-T351 by controlling the welding speed and heat input. *J. Manuf. Process.* **2019**, *46*, 147–158. [CrossRef]
12. Coniglio, N.; Cross, C.E. Mechanisms for solidification crack initiation and growth in aluminum welding. *Metall. Mater. Trans. A* **2009**, *40*, 2718–2728. [CrossRef]
13. Wei, Y.H.; Liu, R.P.; Dong, Z.J. Software package for simulation and prediction of welding solidification cracks. *Sci. Technol. Weld. Join.* **2003**, *8*, 325–333. [CrossRef]
14. Bordreuil, C.; Niel, A. Modelling of hot cracking in welding with a cellular automaton combined with an intergranular fluid flow model. *Comp. Mater. Sci.* **2014**, *82*, 442–450. [CrossRef]
15. Agarwal, G.; Kumar, A.; Richardson, I.M.; Hermans, M.J.M. Evaluation of solidification cracking susceptibility during laser welding in advanced high strength automotive steels. *Mater. Des.* **2019**, *183*, 108104. [CrossRef]
16. Jiang, W.C.; Liu, Z.B.; Gong, J.M.; Tu, S.T. Numerical simulation to study the effect of repair width on residual stresses of a stainless steel clad plate. *Int. J. Pres. Ves. Pip.* **2010**, *87*, 457–463. [CrossRef]

17. Hyde, T.H.; Saber, M.; Sun, W. Creep crack growth data and prediction for a P91 weld at 650 °C. *Int. J. Pres. Ves. Pip.* **2010**, *87*, 721–729. [CrossRef]

18. Pandey, C.; Saini, N.; Mahapatra, M.M.; Kumar, P. Hydrogen induced cold cracking of creep resistant ferritic P91 steel for different diffusible hydrogen levels in deposited metal. *Int. J. Hydrogen. Energy* **2016**, *41*, 17695–17712. [CrossRef]

19. Zhang, Y.J.; Yang, K.; Zhao, J.P. Experimental research and numerical simulation of weld repair with high energy spark deposition method. *Metals* **2020**, *10*, 980. [CrossRef]

20. He, K.F.; Yang, Q.; Xiao, D.M.; Li, X.J. Analysis of thermo-elastic fracture problem during aluminium alloy MIG welding using the extended finite element method. *Appl. Sci.* **2017**, *7*, 69. [CrossRef]

21. Li, H.; Li, J.S.; Yuan, H. A review of the extended finite element method on macrocrack and microcrack growth simulations. *Theor. Appl. Fract. Mech.* **2018**, *97*, 236–249. [CrossRef]

22. Belytschko, T.; Black, T. Elastic crack growth in finite elements with minimal remeshing. *Int. J. Numer. Meth. Eng.* **1999**, *45*, 601–620. [CrossRef]

23. Moës, N.; Dolbow, J.; Belytschko, T. A finite element method for crack growth without remeshing. *Int. J. Numer. Meth. Eng.* **1999**, *46*, 131–150. [CrossRef]

24. Daux, C.; Moës, N.; Dolbow, J.; Sukumar, N.; Belytschko, T. Arbitrary branched and intersecting cracks with the extended finite element method. *Int. J. Numer. Meth. Eng.* **2000**, *48*, 1741–1760. [CrossRef]

25. Chessa, J.; Wang, H.W.; Belytschko, T. On the construction of blending elements for local partition of unity enriched finite elements. *Int. J. Numer. Meth. Eng.* **2003**, *57*, 1015–1038. [CrossRef]

26. Stolarska, M.; Chopp, D.L.; Moës, N.; Belytschko, T. Modelling crack growth by level sets in the extended finite element method. *Int. J. Numer. Meth. Eng.* **2001**, *51*, 943–960. [CrossRef]

27. Song, J.-H.; Areias, P.M.A.; Belytschko, T. A method for dynamic crack and shear band propagation with phantom nodes. *Int. J. Numer. Meth. Eng.* **2006**, *67*, 868–893. [CrossRef]

28. Yang, J.P.; Guo, J.; Qiao, Y.X. *DL/T 869-2012 The Code of Welding for Power Plant*; China Electric Power Press: Beijing, China, 2012. (In Chinese)

29. Yaghi, A.H.; Hyde, T.H.; Becker, A.A.; Williams, J.A.; Sun, W. Residual stress simulation in welded sections of P91 pipes. *J. Mater. Process. Tech.* **2005**, *167*, 480–487. [CrossRef]

30. Arora, H.; Singh, R.; Brar, G.S. Thermal and structural modelling of arc welding processes: A literature review. *Meas. Control.* **2019**, *52*, 955–969. [CrossRef]

31. Zhang, W.Y. *Principle and Technology of Metal*; Fusion Welding; China Machine Press: Beijing, China, 1980. (In Chinese)

32. Zhang, J.X.; Liu, C. *Finite Element Calculation of Welding Stress and Deformation and Its Engineering Application*; Science Press: Beijing, China, 2015. (In Chinese)

33. Asferg, J.L.; Poulsen, P.N.; Nielsen, L.O. A direct XFEM formulation for modeling of cohesive crack growth in concrete. *Comput. Concr.* **2007**, *4*, 83–100. [CrossRef]

34. Benzeggagh, M.L.; Kenane, M. Measurement of mixed-mode delamination fracture toughness of unidirectional glass/epoxy composites with mixed-mode bending apparatus. *Compos. Sci. Technol.* **1996**, *56*, 439–449. [CrossRef]

35. Pan, J.Z. *Practical Manual for Pressure Vessel Materials—Carbon Steel and Alloy Steel*; Chemical Industry Press: Beijing, China, 2000. (In Chinese)

36. Dutt, B.S.; Babu, M.N.; Shanthi, G.; Moitra, A.; Sasikala, G. Investigation on fracture behavior of Grade 91 steel at 300–550 °C. *J. Mater. Eng. Perform.* **2018**, *27*, 6577–6584. [CrossRef]

37. ASME. *ASME Boiler & Pressure Vessel Code, Section II, Part A*; American Society of Mechanical Engineers: New York, NY, USA, 2019.

38. ASME. *ASME Boiler & Pressure Vessel Code, Section II, Part C*; American Society of Mechanical Engineers: New York, NY, USA, 2019.

39. Chen, H.N.; Li, R.F.; Chen, J. *GB/T 24179-2009 Metallic Materials-Residual Stress Determination—The Indentation Strain-Gage Method*; China Standard Press: Beijing, China, 2009. (In Chinese)

Review

Dirlik and Tovo-Benasciutti Spectral Methods in Vibration Fatigue: A Review with a Historical Perspective

Turan Dirlik [1],* and **Denis Benasciutti** [2],*

1   Dirlik Controls Ltd., 12 Main Street, Rugby CV22 7NP, UK
2   Department of Engineering, University of Ferrara, Via Saragat 1, 44122 Ferrara, Italy
*   Correspondence: turan@dirlik.co.uk (T.D.); denis.benasciutti@unife.it (D.B.)

Abstract: The frequency domain techniques (also known as "spectral methods") prove significantly more efficient than the time domain fatigue life calculations, especially when they are used in conjunction with finite element analysis. Frequency domain methods are now well established, and suitable commercial software is commonly available. Among the existing techniques, the methods by Dirlik and by Tovo–Benasciutti (TB) have become the most used. This study presents the historical background and the motivation behind the development of these two spectral methods, by also emphasizing their application and possible limitations. It further presents a brief review of the other spectral methods available for cycle counting directly from the power spectral density of the random loading. Finally, some ideas for future work are suggested.

Keywords: random loading; fatigue damage; power spectral density (PSD); spectral methods

Citation: Dirlik, T.; Benasciutti, D. Dirlik and Tovo-Benasciutti Spectral Methods in Vibration Fatigue: A Review with a Historical Perspective. *Metals* **2021**, *11*, 1333. https://doi.org/10.3390/met11091333

Academic Editor: Anders E. W. Jarfors

Received: 3 August 2021
Accepted: 20 August 2021
Published: 24 August 2021

Publisher's Note: MDPI stays neutral with regard to jurisdictional claims in published maps and institutional affiliations.

1. Introduction

The estimation of fatigue life under variable amplitude or random loadings has been an active research topic in the last fifty years, and the activity has further increased in the last two decades. Cumulative damage calculations, in particular, occupy a dominant sector in the structural integrity assessment of metallic components and structures subjected to random fatigue loadings. In order to achieve a high level of structural reliability, fatigue life calculations must be made at several stages of the design and development process. An important aspect of the development of fatigue-resistant components and structures is the ability to estimate component fatigue life and, thus, prevent unexpected failures.

The traditional approach to fatigue analysis—often referred to as "time domain approach"—uses a technique called rainflow cycle counting to decompose a variable amplitude time signal of stress into fatigue cycles [1,2]. The damage from each cycle is then computed using an S/N curve, which characterizes the material strength for constant amplitude loadings. The damage over the entire time signal is finally calculated by summing the damage from all the individual cycles, using, for example, the celebrated Palmgren–Miner linear damage rule. This simple rule, which sums the damage of cycles, regardless of their order in the loading, also postulates that fatigue failure occurs when the damage exceeds a critical level equal to unity [3].

The time domain approach does not present any particular theoretical complexity and, nowadays, it is easy to implement by considering, for example, that the rainflow algorithm is available in some numerical packages [4,5]. The approach may, however, require a long computation time when it has to analyze a lot of digitalized signals of long duration, such as those computed in hundreds of thousands of nodes in a finite element model. This unfortunate circumstance could render the fatigue damage computation the bottleneck of the whole design process.

In addition, the procedure by which stress signals are obtained plays an important role. In fact, the computation time may increase if the random stress signals are obtained by a transient simulation in time domain, which is orders of magnitude slower than an analysis in frequency domain, able to compute harmonic transfer functions and stress power spectral densities with much less computational effort.

For linear systems, fatigue life calculations conducted entirely in the frequency domain can shorten the computation time considerably. This corresponds to the so-called "frequency domain approach" in vibration fatigue. Fatigue life calculation methods based on frequency domain information of a random loading are now well established, and suitable commercial software is commonly available (e.g., nCode DesignLife, Simulia Fe-safe, CAE-fatigue, Ansys Random Vibration Fatigue). Among frequency domain methods, one should also number those approaches in which spectral methods are combined with crack growth models with the aim to estimate fatigue life under random loadings [6–10]. Crack growth models, though relevant for the fatigue durability field, are not included in the following discussion, which is concerned with S/N based approaches in frequency domain.

Since the early works of Miles [11] and Bendat [12], and in particular in the past thirty years, tens of spectral methods have been developed and their number continues to rise even today (a short list is found in [13,14]).

However, among all the existing methods, two of them have gained an increased popularity and large use in the scientific and technical community: they are the Dirlik method [15] and the Tovo–Benasciutti (or TB) method [16,17]. One reason for their popularity is their accuracy, often superior to other methods, as highlighted by several comparative studies that are commented on in the following sections.

Among the multitude of spectral methods, the Dirlik and TB methods are specifically considered hereafter as they were invented by the two authors of this paper, who thus have a vantage point in explaining the motivation behind the development of both methods, meanwhile emphasizing some of their theoretical peculiarities. The purpose of this paper is also to comment on their progresses and possible shortcomings, as well as to discuss their relationship with a few other spectral methods, providing some insights into future work. To this regard, this paper is not meant to be, in the strict sense, either a critical review or a comparison of lots of spectral methods; rather, its aim is to focus on the Dirlik and TB methods while also casting a glance on how they interrelate with other existing methods.

2. Random Processes in Frequency Domain: Spectral Properties and Fatigue Damage

2.1. Spectral Properties

Fatigue life assessment is critical in the design of components and structures exposed to random loads, such as a car on a rough road or a wind turbine. These loads can be viewed as the realization of a stationary random process $s(t)$, which can be described in the frequency domain by a power spectral density (PSD) function $G(\omega)$, where ω is the circular frequency (in rad/s). The PSD is related to the Fourier transform of $s(t)$ [18]. It provides a picture of the energy content of $s(t)$ over frequencies. The power spectrum is often used to describe signals for qualification tests [19]. Note that two definitions of PSD exist, namely single-sided versus double-sided. Each one can be function of circular frequency ω (rad/s) or frequency $f = \omega/2\pi$ (Hertz). $G(\omega)$ is a single-sided PSD function of ω.

It is customary to classify a random signal $s(t)$ based on its frequency content, that is, on the shape of its PSD. The signal is said to be "narrow band" if $G(\omega)$ has a peak around a single frequency, generally the resonant frequency of a vibrating system. In all other cases in which the PSD covers a wider range of frequencies, the random signal is not narrow band, and it is named as "wide band" (or "broad band"). Sometimes, more specific definitions ("bimodal", "trimodal", "multimodal", etc.) are adopted to specify that a PSD has two, three, or more well-defined peaks.

Figure 1 compares three time history samples belonging to three types of random processes. The figure emphasizes quite well the differences among the various time histories based on their corresponding PSD, although it considers power spectra with rectangular blocks that are only idealizations of the smoother spectra normally encountered in practical applications.

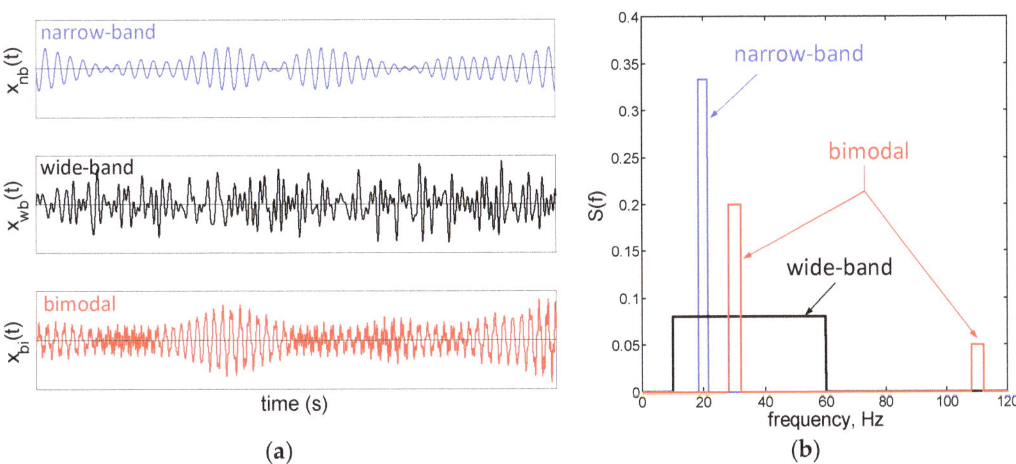

Figure 1. (**a**) Time history samples and (**b**) their PSD (narrow band, wide band, bimodal).

The definitions of narrow band or wide band process introduced so far are merely qualitative. A more quantitative definition is provided by means of several types of spectral parameters, which are indeed introduced with the purpose of establishing to which degree a random signal is narrow band or wide band. The first parameters are the spectral moments [18]:

$$m_n = \int_0^\infty \omega^n G(\omega)\, \mathrm{d}\omega, \quad n = 0, 1, 2 \ldots \tag{1}$$

The theory shows [18] that, for a random loading with zero mean value, the spectral moment of order zero coincides with the variance of the random signal, $m_0 = Var(s(t))$; the root mean square (rms) value is $\sigma_x = \sqrt{m_0}$.

Other spectral moments of higher order—or, better, their combinations—are also used to characterize some statistical properties of the random signal that are of interest in fatigue analysis. For example, for a Gaussian process, $\lambda_0^+ = \sqrt{m_2/m_0}$ (rad/s) is the expected number of mean value upcrossings per unit time, while $\mu = \sqrt{m_4/m_2}$ (rad/s) is the expected number of peaks per unit time [18]. For what follows, it is useful to introduce the "mean frequency" $\omega_1 = m_1/m_0$ (rad/s) of the spectral density; it may be interpreted as the distance of the centroid of the spectral mass from the frequency original [20].

Besides the previous parameters, other quantities, known as bandwidth parameters, are used extensively. They are combinations of spectral moments. Various equivalent definitions exist, but those most used are [18]:

$$\alpha_1 = \frac{m_1}{\sqrt{m_0 m_2}}, \quad \alpha_2 = \frac{m_2}{\sqrt{m_0 m_4}}, \quad \alpha_{0.75} = \frac{m_{0.75}}{\sqrt{m_0 m_{1.5}}} \tag{2}$$

They belong to a more general class, $\alpha_k = \frac{m_k}{\sqrt{m_0 m_{2k}}}$. Note the relationship with the Vanmarcke's spectral parameter $q = \sqrt{1 - \alpha_2^2}$ [20]. Both α_1 and α_2 approach unity for a narrow band process, whilst they approach zero when the spectral width increases. Parameter α_1 is related to the definition and properties of the envelope of the random process, and it is often called "groupness parameter" in ocean engineering. Parameter α_2 is often called the "irregularity factor", which is a useful parameter characterizing the behavior of the process. In time domain, the irregularity factor is defined as the ratio of the number of mean value upcrossings n_0^+ to the expected number of peaks n_p, that is $IF = n_0^+/n_p$; for a Gaussian process it takes the form $\gamma = \lambda_0^+/\mu$, which coincides with α_2. This parameter lies between zero and unity. It approaches unity for a narrow band process in which peaks and troughs are placed symmetrically above and below the mean value of the process. It approaches zero when the process has many peaks between two successive crossings of its mean value.

2.2. Fatigue Damage

The fatigue damage of a random time history $s(t)$ of time length T is computed in two stages. In the first, a counting method (e.g., rainflow) is applied to $s(t)$, with the purpose of identifying the set of fatigue cycles [2]. Counted cycles have, in general, different amplitudes and mean values, and thus they contribute with a different damage. Therefore, in the second stage, the damage of every counted cycle is summed up to determine the damage of $s(t)$ as a whole. This step calls for a damage accumulation rule, as the Palmgren–Miner linear law [3]:

$$D = \sum_i^{N(T)} \frac{n_i}{N_i} \tag{3}$$

This summation extends over the total number of cycles $N(T)$ counted in $s(t)$; n_i numbers the cycles with amplitude s_i, which would cause a failure after N_i repetitions in a constant amplitude test.

Often, the relationship between stress amplitude s_i and cycles to failure N_i is expressed by a straight line in a log–log diagram, through the S/N curve $Ns^b = K$. This equation is fitted to experimental data by regression analysis. In some cases, a two-slope equation is preferred when it fits experiments better.

In Palmgren–Miner rule, fatigue failure is predicted to occur when damage D reaches a critical value equal to unity, even though lower values are recommended in some design codes (e.g., [21]) to account for experimental evidence.

The summation in Equation (3) assumes that the cycles counted in $s(t)$ are grouped into bins of n_i cycles, all having the same amplitude s_i. The actual values of n_i and s_i obviously depend on the actual course of $s(t)$. If the instantaneous values of $s(t)$ vary randomly—that is, $s(t)$ is modeled as a random process—it follows that n_i and s_i are both random variables. In this case, the summation in Equation (3) needs to be reformulated in a probabilistic way:

$$D = n_{rfc} \int_0^{+\infty} \frac{1}{N(s)} p(s) ds \tag{4}$$

Here, $p(s)$ is the probability distribution of the amplitudes of rainflow cycles, n_{rfc} is the number of rainflow cycles counted in T, and $N(s)$ is the number of cycles to failure at constant amplitude s. Notice that the previous expression is very general, as $N(s)$ is not restricted to a straight-line equation.

Although, at first glance, the previous formula may appear complicated, it is nothing more than the Palmgren–Miner rule in Equation (3) extended to continuously distributed amplitudes.

Equation (4) can be solved in closed form if one knows the analytical expressions of $p(s)$. This circumstance occurs, for example, when $s(t)$ is a narrow band process. In this case, peaks and valleys are placed symmetrically around the mean value of $s(t)$, and each cycle is formed by pairing each peak to the adjacent valley (this corresponds to the peak counting method)—the range is the peak–valley distance. Consequently, each cycle has zero mean and amplitude coincident with the peak value, which in a narrow band process follows a Rayleigh distribution [18]. In a narrow band process, the number of cycles counted is also known, it being equal to the number of crossings of the mean value, $n_{\mathrm{rfc}} = \lambda_0^+ T$. When inserted into Equation (4), the previous mathematical conditions yield the expression of the "narrow band" damage in time interval T [18]:

$$D_{NB} = \frac{\lambda_0^+ T}{K} \left(\sqrt{2m_0} \right)^b \Gamma \left(1 + \frac{b}{2} \right) \tag{5}$$

in which $\Gamma(-)$ is the gamma function. This expression is usually credited to Miles [11] or Bendat [12]. It is restricted to a straight S/N line. In case the S/N relationship is not straight, it is possible to solve Equation (4) numerically, by taking $p(s)$ as a Rayleigh distribution.

When the random process is no longer narrow band, the previous formula becomes too conservative. Indeed, in a wide band process, fatigue cycles cannot simply be obtained by joining a peak with a symmetric valley, as is done in the peak counting [2]; actually, this counting procedure would yield cycles with amplitudes larger than those identified by the rainflow counting. On the other hand, the algorithm of the rainflow counting is so intricate that it seems not possible to determine, in closed form, the expression of the probability distribution of amplitudes and mean values of rainflow cycles, and to relate it to the bandwidth parameters of the wide band process.

For this reason, historically, the approach followed initially—easy but oversimplified—was to introduce a correction factor less than unity, by which to reduce the narrow band damage in Equation (5). The correction factor must become smaller and smaller when the spectral bandwidth of the random process becomes wider and wider. Since this inverse relationship is not known, and it cannot be determined in closed form because of the same reasons mentioned above for the rainflow counting, the only feasible way was to calibrate the damage correction factor based on simulation results. The difficult task was to select which bandwidth parameters must enter the correction factor. Assumptions were made based on the trends observed in simulations.

Among the spectral methods that followed this approach, the first, and probably the most famous, one is that of Wirsching and Light, proposed in 1980 [22], in which the correction factor is made to depend solely on the spectral parameter α_2. In subsequent years, other methods following the same idea were proposed. In those methods, the expression of the correction factors was refined by introducing the dependency on other bandwidth parameters. For example, an approach named the "empirical $\alpha_{0.75}$-method" proposed a correction factor in the form $\alpha_{0.75}^2$ [17].

It should come as no surprise that the use of a correction factor for estimating the damage of a wide band process, no matter how simple, has the drawback of not providing the amplitude probability distribution $p(s)$ of the cycles causing that damage. Knowing $p(s)$ indeed has several advantages. First, it makes it possible to compute the fatigue damage not only for a straight S/N line, but also for any smooth curve with continuous change of slope, with or without endurance limit. Second, the amplitude distribution allows one to determine the cumulative (or loading) spectra of rainflow cycles, and to extrapolate it towards cycles with large amplitudes (rare events).

Both Dirlik and TB methods illustrated below belong to the category of approaches that provide an estimate of the amplitude probability distribution of rainflow cycles. For a more comprehensive survey on spectral methods, the reader may refer to [13,14].

3. Dirlik Method (1985)

This method can, for sure, be numbered as a milestone among spectral methods. This is testified by the number of citations, to date more than 600 in Google Scholar [23].

As already said, a power spectral density function specifies how the characteristics of a random signal are distributed over a given frequency range. This information is needed in many engineering sectors; techniques to calculate the PSD from a time record are readily available in the form of fast Fourier transform (FFT).

Early in the development of the Dirlik method, because of the complexity of the rainflow counting algorithm, it was decided that it would be such a difficult task to derive in closed form the distribution of rainflow cycles from the PSD function $G(\omega)$. A Monte Carlo approach was thus employed to generate a sample stress history $s(t)$ from $G(\omega)$ using FFT methods [24,25]. Then, the rainflow algorithm was used on $s(t)$ to extract the cycles and the probability density function of rainflow counted ranges, which in turn allowed calculating the fatigue damage for any given material constants in an S/N curve.

The idea of the Monte Carlo approach is based on the "weak law of large numbers", which states that the sample average converges in probability towards the expected value if the sample size increases [26]. It is then important that, in numerical simulations, the sample stress histories are sufficiently long to contain a large number of turning points (peaks and troughs); in fact, the longer the time history, the more chance of catching rare events.

The exact procedure of the approach linking PSDs to rainflow range probability densities is explained fully in [15,27], and it is briefly summarized here.

A total of 70 PSDs of various shapes were considered, some rectangular and some smoothly varying, as shown in Figure 2. To make the comparison easier, power spectra were normalized so that all had the same rms value σ_x and the same expected rate of peaks μ (that is, the same number of rainflow cycles counted in a time interval). With these two parameters kept fixed, the irregularity factor γ was made to vary (from 0.16 to 0.99) by adjusting the parameters defining the power spectrum shape—which in turn corresponds to varying the spectral moments of order zero, two, and four (m_0, m_2, m_4).

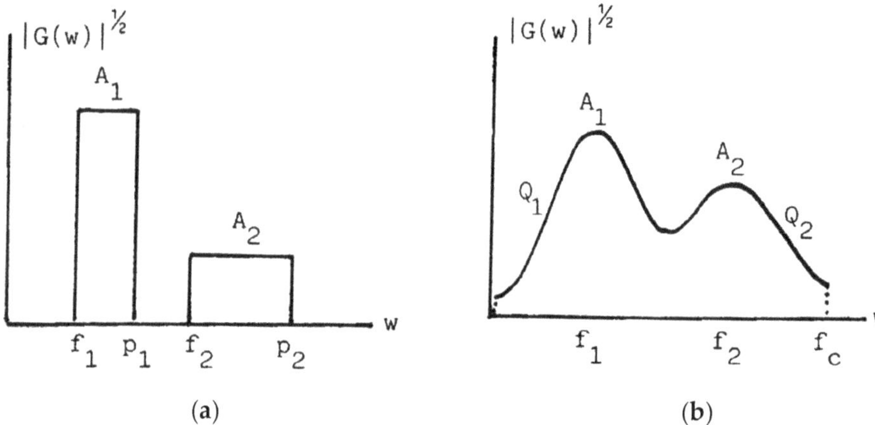

Figure 2. Types of PSD used in numerical simulations used to develop the Dirlik method: (**a**) rectangular bimodal spectrum; (**b**) smooth spectrum. Figures are from [15].

The rectangular bimodal spectrum (see Figure 2a) was used extensively because of its simplicity to assume a wide range of values of the irregularity factor γ, having the same rms value σ_x and the same expected rate of peaks μ, by adjusting the amplitudes and the frequency boundaries of the power spectrum.

In preliminary simulation trials, it was, however, observed that the mean frequency ω_1 of the spectrum (that is, its first order moment m_1) also has a role in changing the fatigue damage of a simulated time history. To investigate this relationship more closely, 42 out of 70 power spectra were shaped so as to have the same σ_x, μ, γ, but different m_1. Because the expected rate of peaks μ is identical for all spectra, a "normalized mean frequency" $x_m = \omega_1/\mu$ was introduced:

$$x_m = \frac{m_1}{m_0}\sqrt{\frac{m_2}{m_4}} \tag{6}$$

For a given power spectral density, sample stress time histories $s(t)$ were generated using the inverse FFT (IFFT). A single generated stress time history had 1024 points. The simulation procedure was repeated 20 times in order to obtain a sufficiently long record of stress time history. This long record (called one block) consisted of $1024 \times 20 = 20{,}480$ time points [15]. The number of 1024 FFT points was not a choice, but it was dictated by the maximum size of the memory and disk capacity of computers available at that time. The use of 20 replicated stress time histories had the very purpose of increasing the entire length of the simulated block. Processed in time domain, the block was converted into a sequence of peaks and troughs, from which fatigue cycles were extracted by means of the rainflow counting and the simple-range counting (which pairs a peak with the next valley [2]). From the set of counted cycles, the probability distributions of simple ranges and rainflow ranges were finally calculated. An example is displayed in Figure 3.

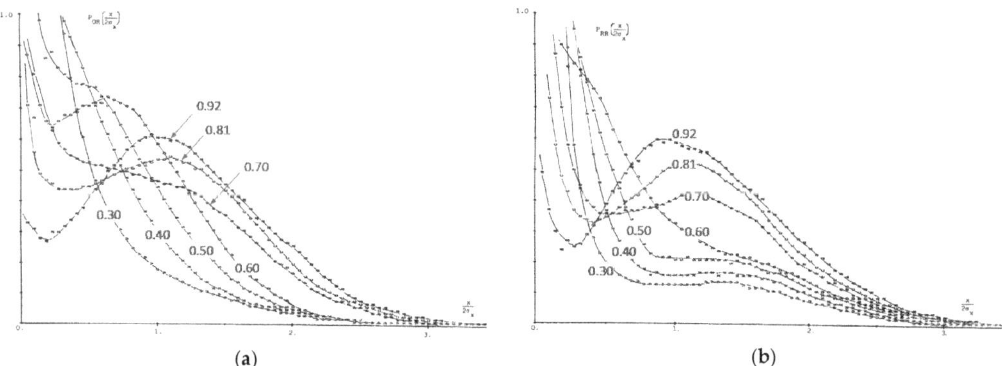

(a)   (b)

Figure 3. Probability distribution of: (**a**) simple ranges; (**b**) rainflow ranges. The numbers correspond to the irregularity factor of smooth PSD. On the x-axis, values are normalized as $x/2\sigma_x$ [15].

Thanks to the simulation results, it was argued that the probability density function (PDF) of rainflow ranges is to be a mixture of three distributions: an exponential function, a Rayleigh function with variable parameter, and a standard Rayleigh function. The full expression in terms of a normalized variable $Z = \frac{r}{2\sqrt{m_0}}$ is [15]:

$$p_{DK}(Z) = \frac{1}{2\sqrt{m_0}}\left[\frac{D_1}{Q}e^{-\frac{Z}{Q}} + \frac{D_2 Z}{R^2}e^{-\frac{Z^2}{2R^2}} + D_3 Z e^{-\frac{Z^2}{2}}\right] \tag{7}$$

where r is the rainflow range. The coefficients D_1, D_2, D_3 and R are defined as

$$D_1 = \frac{2\left(x_m - \alpha_2^2\right)}{1+\alpha_2^2} \quad D_2 = \frac{1-\alpha_2-D_1+D_1^2}{1-R} \quad D_3 = 1 - D_1 - D_2,$$
$$R = \frac{\alpha_2 - x_m - D_1^2}{1-\alpha_2-D_1+D_1^2} \quad Q = 1.25\frac{(\alpha_2-D_3-D_2 R)}{D_1} \tag{8}$$

where x_m is the normalized mean frequency in Equation (6).

Note that the symbols in Equations (7) and (8) conform to the notation used nowadays, which differs from that originally used in [15]. The quantities D_1, D_2, D_3, and R are "best fit" parameters that turned out after minimizing the difference between the observed probability distribution and the analytical one.

The probability distribution in Equation (7) represents the link between the rainflow counted ranges and the power spectral density. The importance of this equation lies in the fact that, once it is used to determine the PDF of rainflow ranges, the life estimation can be made with any form of S/N data by using Equation (4). The use of this method is not restricted to a straight line representation of S/N data on a log–log scale only, which instead could be a smooth curve with continuous change of slope with or without endurance limit. In the case of single slope S/N line, $Ns^b = K$, the substitution of $p_{DK}(Z)$ into Equation (4) returns a closed-form expression for the damage in time interval T [17,28]:

$$D_{DK} = \frac{\nu_p T}{K}(\sqrt{m_0})^b\left[D_1 Q^b \Gamma(1+b) + \left(\sqrt{2}\right)^b \Gamma\left(1+\frac{b}{2}\right)\left(D_2|R|^b + D_3\right)\right] \tag{9}$$

That described so far is the original version Dirlik method commonly used. A temperature-modified version was also proposed in [29–31] to estimate the high-cycle fatigue damage for uniaxial loadings caused by random vibrations directly from a power spectral analysis. The model combines the frequency-based method and the temperature effect, and it is verified by comparison with experimental test results for a high-pressure die-cast aluminum alloy. Actually, the approach in [29] incorporates the temperature effect into the Dirlik method by considering a temperature-dependent inverse slope and fatigue strength $b(T)$ and $K(T)$, and by also taking the temperature time history into consideration by using a weighted sum of the damage at each temperature. In this way, the temperature-dependent spectral approach, though developed in [29] only for the Dirlik method, is in fact applicable to any other spectral method, provided that temperature-dependent S/N parameters are used.

4. Tovo–Benasciutti (TB) Method (2002, 2005)

The theory of this method was first laid out in [32]. It postulates that the amplitude–mean joint probability distribution of rainflow cycles lies between two limit distributions, and can be estimated as their linear combination:

$$p_{TB}(s,m) = w\, p_{LCC}(s,m) + (1-w)\, p_{RC}(s,m) \tag{10}$$

Here, w is a weight factor that needs to be determined. Unlike Dirlik method, the TB method provides the joint distribution of amplitudes and mean values of rainflow cycles. The two functions $p_{LCC}(s,m)$ and $p_{RC}(s,m)$ represent the amplitude–mean distributions of the level-crossing counting (LCC) and of the simple-range counting (RC)—actually, the latter is only approximated. Their analytical expressions, not reported here, can be found in [16,32]. The two distributions are shown in Figure 4. Note that $p_{LCC}(s,m)$ involves two Dirac delta functions.

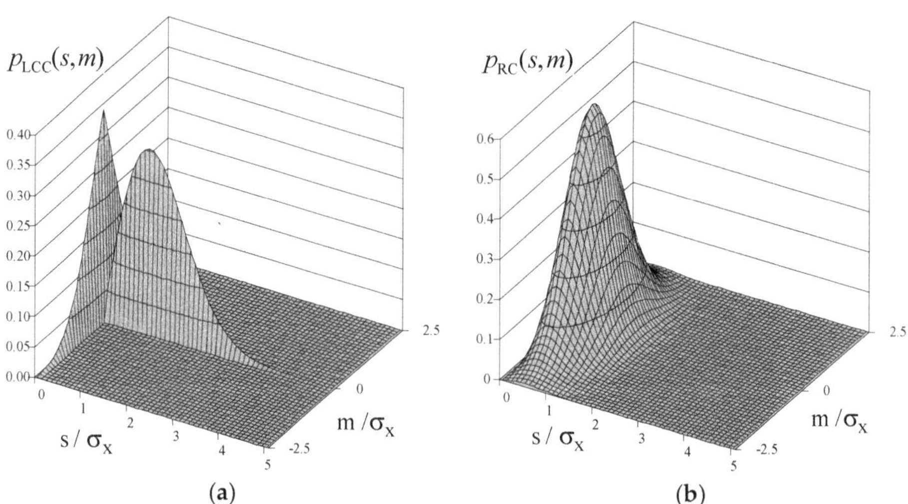

Figure 4. Amplitude–mean probability distributions in TB method: (**a**) level crossing counting; (**b**) approximate range counting. Representation of $p_{LCC}(s,m)$ is only qualitative, as it includes a Dirac delta function for $s=0$ and $m=0$. Reproduced from [32] with permission from Elsevier.

Equation (10) shows that the LCC and RC distributions represent two bounds between which the rainflow cycle distribution is enclosed. In other words, Equation (10) postulates that the rainflow cycle distribution of any random process lies between two limit distributions, and its actual shape depends on the value of w.

Likewise in Equation (10), the same weighted sum holds also for the marginal probability distributions $p_{TB}(s)$, $p_{LCC}(s)$, $p_{RC}(s)$, and, most importantly, for the damage values:

$$D_{TB} = w\, D_{LCC} + (1-w)\, D_{RC} \cong \left[w + (1-w)\alpha_2^{b-1} \right] D_{NB} \tag{11}$$

The latter inequality takes advantage of the fact that $D_{LCC} = D_{NB}$ and that the simple-range counting damage is approximated as $D_{RC} \cong \alpha_2^{b-1} D_{NB}$ [16,17]. Though, from a practical point of view, the quantity $\left[w + (1-w)\alpha_2^{b-1} \right]$ can be interpreted as a correction factor of the narrow band damage, the above arguments clearly demonstrate that the origin of this factor is in fact quite different, and the TB method has a sound theoretical basis.

Turning back now to the weighted sum that links probability distributions and damage values through w, the next step to complete the definition of the TB method was to find a proper expression for w. At least theoretically, parameter w is a function of the whole set of spectral and bandwidth parameters of the PSD of the random process.

Nevertheless, it was—as it seems even today—rather hard, if not impossible, to demonstrate theoretically which spectral parameters contribute to the definition of w. Some trends can be argued. In the narrow band case, for example, the rainflow distribution must converge to the Rayleigh one (i.e., LCC function), which in turn requires that $w \cong 1$ for a narrow band process in which α_1 and α_2 both approach unity. Other trends emerged from numerical simulations. The approach was to generate stress time histories, as in the procedure described for the Dirlik method.

Using simulation results with power spectra densities of various shapes, a first expression was proposed in [32]:

$$w_1 = min\left(\frac{\alpha_1 - \alpha_2}{1 - \alpha_1}, 1 \right) \tag{12}$$

This formula is yet not continuous. Moreover, its accuracy tends to diminish if α_1 and α_2 differ significantly [13], as in the special case of a lightly damped linear oscillator driven by white noise, whose response has $\alpha_1 \cong 1$ but $\alpha_2 = 0$ (irregular process) [18].

From a broader perspective, conflicting views seem to emerge in the literature regarding the accuracy of parameter w_1. While the agreement with simulation results was judged as "excellent" in [32], a less flattering opinion—moreover consistent with the results of [33]—is given in [34], where the use of this formulation of w_1 is indeed discouraged.

Apart from this, long before the above studies, it was soon discovered that Equation (12) could be ameliorated by seeking better solutions. Although a first attempt was made in [35] to find a continuous but equally accurate expression to replace Equation (12), it was only in [36], and later in [16], that the expression of w used today has finally been developed:

$$w_2 = \frac{(\alpha_1 - \alpha_2)\left[1.112(1 + \alpha_1\alpha_2 - (\alpha_1 + \alpha_2))\, e^{2.11\alpha_2} + (\alpha_1 - \alpha_2) \right]}{(1 - \alpha_2)^2} \tag{13}$$

This expression is a result of a best fitting on simulation results from power spectral densities of various shapes (see Figure 5). Similar to the Dirlik method, this formula assumes that the rainflow probability distribution is linked to four spectral moments, m_0, m_1, m_2, and m_4. The soundness of this correlation, first foreseen by Dirlik, was confirmed after comparing frequency domain estimations with time domain simulation results. It was shown in [16] that the use of Equation (13) along with Equation (11) guarantees an excellent accuracy.

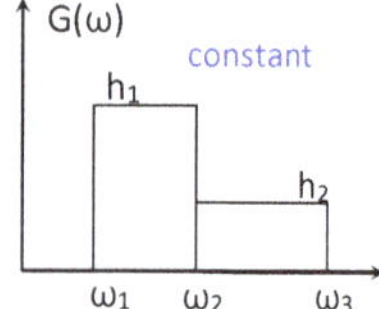
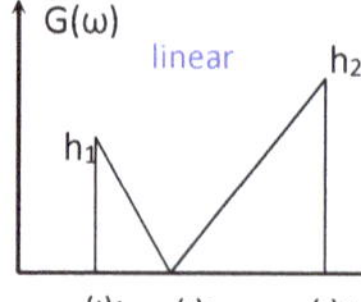
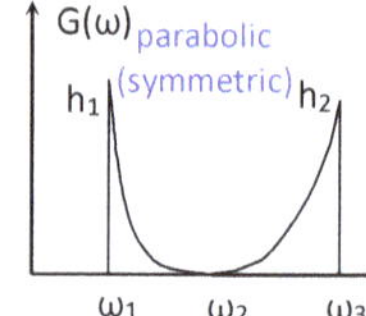
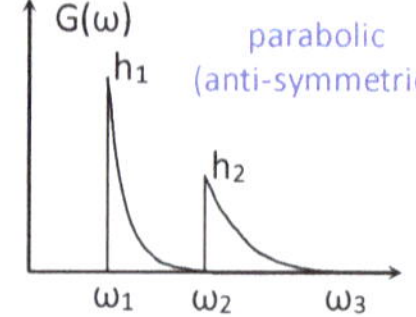
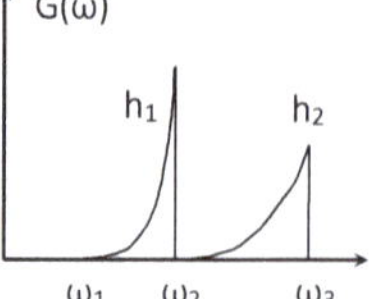

Figure 5. Power spectral densities used in simulations to calibrate weight w_2 in TB method.

After that, this accuracy was further confirmed by other studies (e.g., [33,37]), and the TB method rapidly became popular and widely used, similar to the Dirlik method—though the latter seems to remain the method preferred by most scholars.

Soon after [16] was published, a further attempt was made in [17] to correlate w to fractional order spectral moments $m_{0.75}$ and $m_{1.5}$ (in bandwidth parameter $\alpha_{0.75}$) by the formula

$$w_3 = \frac{\alpha_{0.75}^2 - \alpha_2^2}{1 - \alpha_2^2} \tag{14}$$

This choice was suggested after having observed that fractional spectral moments were considered also by other spectral methods (Zhao-Baker [38], "single-moment" [39,40], "empirical $\alpha_{0.75}$-method" [17]). Against all odds, the comparison made in [17] even revealed that the accuracy of Equation (14) is slightly better than Equation (13); however, despite this outcome, quite surprisingly, Equation (14) has been ignored by scholars in favor of Equation (13).

5. Area of Application of Spectral Methods

Since their early development, but especially from the appearance of Dirlik approach in 1985, spectral methods have found ever more application in many engineering fields, even those much different from each other. Over the years, many examples and case studies have been developed in engineering sectors such as automotive, wind, offshore, and marine, not to mention the field of aerospace and advanced composite materials.

Obviously, as already pointed out for spectral methods, here it is also not possible (as well not very useful) to enumerate all the examples or case studies cited in the literature, by also considering that probably a great many of them were not published as scientific articles. Nor is the aim of this section to present a summary list of articles. Rather, the lines that follow have the only purpose of casting a glance at a few representative applications of spectral methods, with special interest on examples that analyze real structures, often with the aid of finite element simulation.

It has indeed to be pointed out that one of the main advantages of spectral methods (which may in fact explain their widespread use) is the possibility to use them in conjunction with computer-aided analysis. Typically, once a finite element model is built and subjected to a known random acceleration/load expressed by a power spectral density, the stress power spectra in each node are calculated and then input directly in a spectral method, which then provides a damage value in each node of the model. If the nodal stress is multiaxial (i.e., a PSD matrix is defined), a multiaxial criterion must be used [41].

Among those existing, multiaxial spectral methods that transform a multiaxial stress into an equivalent uniaxial stress permit the "traditional" spectral methods (like as Dirlik, TB method or others) to be used [42]. This synergistic combination of multiaxial criteria and frequency domain methods has allowed the field of application of spectral methods to be expanded considerably.

In the automotive field, application of spectral methods generally relies on results of multibody and/or finite element simulation at various levels of modeling detail (chassis, engine, tires, etc.) [43–47]. Sometimes the focus is on a specific component—structural or not—that has to endure random vibrations due to road roughness. Interesting is, for example, the study in [48], which applies spectral methods for estimating the road-induced fatigue damage in a carbon steel coil spring based on acceleration signals acquired in a car driving on various road types [48]. A similar goal is pursued in [49], where field measurements and finite element modeling are used for analyzing a structural component of an automotive headlight. The capabilities of spectral methods as design tool are further demonstrated in [50] by an experimental/numerical analysis of a wheel of an industrial vehicle. When computation time becomes essential, it is possible to integrate spectral methods in an algorithm for monitoring fatigue damage in real time using onboard equipment, as demonstrated by the example in [51] with a heavy-duty truck frame.

In the field of maritime engineering, where there are different types of structures subjected to random waves (e.g., ships, mooring systems, offshore wind turbines, and fixed offshore structures), the use of spectral methods is particularly effective. An advantage is that the power spectra for certain sea states (e.g., JONSWAP, Pierson–Moskowitz) are known in closed form and can be input directly in a finite element analysis aimed at determining the structural dynamic behavior and stress response. Frequency-based fatigue analysis methods are also mentioned in design guidelines [52].

An example of frequency domain fatigue analysis of a semi-submersible platform is described in [53]. The study applies spectral methods with multiaxial fatigue criteria to the output of a three-dimensional finite element model. The case study also serves as a basis to compare the accuracy of various spectral methods.

When data storage and computational cost in recording and processing stress time histories are important, spectral methods prove to be particularly advantageous. For example, [54] proposed a structural monitoring system for a ship hull in which a damage detection algorithm using spectral methods is embedded into a wireless sensor.

For offshore wind turbines, the combined effect of wave and wind loadings acting simultaneously is investigated in [55]; the contribution of different sea and/or wind states can also be taken into account. Spectral methods are also used for estimating the structural integrity of welded joints, which constitute a typical structural detail in offshore structures [55,56].

Examples specific to the field of wind engineering show how spectral methods can be used to predict the fatigue damage of wind-excited structures [57,58]; corrections factors are introduced to account for non-Gaussian effects in the output stress caused by nonlinear aerodynamic damping. A spectral based fatigue analysis, integrated with finite element simulation, is the basis for a structural integrity assessment of two welded joints in a wind turbine tubular tower subjected to different wind speed and direction conditions [59]. When a best balance is required between accuracy and overall computation cost, the study suggests the use of the frequency domain approach over the time-domain one—and among spectral methods, TB and Dirlik methods are recommended [59].

The literature also offers examples of use of spectral methods in engineering sectors such as aerospace and electronics, or with composite materials. In the former cases, for example, spectral methods combined with finite element analysis are employed for the virtual qualification of electronic assemblies subjected to aerospace vibratory environment [60]. Other examples with electronic devices [61,62] are discussed in the following section. Concerning composite materials, spectral based approaches attempt to reformulate strength criteria for composite materials (e.g., Tsai–Hill [63], residual strength, or stiffness model [64,65]) to extend their validity to the frequency domain for the case of random vibrations [66]. Even for metal/composite assemblies undergoing random vibrations, an example demonstrates how to combine finite element analysis with spectral methods successfully [67]. Another application is described in [68] for carbon fiber-reinforced silicon carbide ceramic matrix composite, which are considered as excellent materials for thermal protection structures of launch vehicles and spacecraft structures and, for this reason, exposed to thermo-acoustic vibration loading in service.

Besides the examples above, other, more specific, applications of spectral methods deal with welded [56] or riveted joints [69], to nonferrous materials [70] or even to nonmetallic materials such as concrete [71].

6. Comparison of Spectral Methods

Although the discussion so far has focused on Dirlik and TB approaches, it should not be forgotten that, in the literature, a great many other spectral methods exist for wide band random processes. They are nevertheless too numerous to list them all here. The following list, certainly not exhaustive, mentions in chronological order those spectral methods that seem to be among the most cited in the literature:

- Wirsching and Light (1980) [22].
- Ortiz and Chen (1987) [72].
- "Single-moment" (1990) [39,40].
- Zhao and Baker (1992) [38].
- Steinberg three-band method (2000) [73].
- Fu–Cebon (2000) [74] and Modified Fu–Cebon (2007) [75].
- Empirical $\alpha_{0.75}$ method (2006) [17].
- Gao and Moan (2008) [37].
- Lalanne (2009) [76].

All these frequency domain methods are usually derived for stationary Gaussian processes, but extensions to stationary non-Gaussian processes or to specific subclasses of nonstationary processes (e.g., switching case) have also been proposed. Except for the narrow band case, all these methods are approximate; they differ in what they estimate. While some methods estimate only the expected fatigue damage, others also attempt to also estimate the probability distribution of rainflow amplitudes [17]. Some methods in the previous list (e.g., "single-moment", Fu–Cebon method and its modified version, Gao and Moan) were developed for bimodal or trimodal processes, and they are now considered as a safe technique to assess the vibration fatigue life of real engineering structures.

With the increase in the number of spectral methods, the need has come to establish which of them is "the best" method. A lot of articles published over the last decades have proposed systematic comparisons among spectral methods, and therefore this work does not repeat such a similar analysis. Rather, it summarizes the main findings from other articles. Readers are advised to draw their own conclusions as to which method is "the best" for their own application after investigating each method's assumptions, limitations, and application areas. When interpreting the outcomes of comparison studies, an important aspect to consider is the length of simulated time histories, which should be the longest possible so as to increases the chance of observing large amplitude cycles that appear only occasionally, and thus to make the comparison more reliable.

6.1. Comparison between Dirlik and TB Method

Before comparing these two methods with others, it is useful to give them a closer inspection. To this end, this section compares the Dirlik and TB methods for some selected power spectral densities. The comparison considers not only the amplitude probability distribution, but also the "damage distribution" (to be defined below) and the estimated damage. The purpose is to show that, even though the methods yield, in general, very close predictions, some differences may sometimes occur in their amplitude distributions.

The following discussion deals with the power spectral density in Figure 2a, which is defined by the area ratio $r_A = A_2/A_1$ and by frequencies p_1, f_1, p_2, f_2. The frequency ratios p_1/f_1 and p_2/f_2 can be chosen so as to have different [15] or equal [39,40] values. The condition $c = p_1/f_1 = p_2/f_2$ ensures the same spectral bandwidth for both blocks. In this case, the power spectrum is only defined by ratio of areas $r_A = A_2/A_1$ and frequencies $r_f = f_2/f_1 = p_2/p_1$. This type of power spectrum is particularly advantageous, as it allows not only bimodal cases, but also narrow band and band-limited power spectra to be obtained by a suitable choice of geometric parameters. When $r_A = 0$ (whatever r_f) the spectrum becomes unimodal—whether narrow band or wide band depends on c. A small value (e.g., $c = 1.1/0.9$) assures that each rectangle is narrow band. As c increases, the spectrum width widens. Additionally, for this power spectrum, it is easy to obtain closed-form expressions that express $\alpha_1, \alpha_2, \lambda$ in terms of c, r_B, r_f (see [13]). With such formulae, parameters r_A and r_f can be varied so as to obtained prescribed values of α_1 and α_2.

A short list of PSD geometrical and spectral parameters is summarized in Table 1. For every spectrum, the lowest frequency f_1 is adjusted so that all power spectra have the same number of peaks per second $\mu = 20$. The three columns on the right provide the damage ratio D_{TB}/D_{DK} for three values of S/N slope.

Table 1. Parameters of some types of narrow band, band-limited, and bimodal power spectral densities for which Dirlik and TB methods are compared (f_1 in Hz, μ peaks per second, λ_0^+ crossing per second, n.a. = not applicable).

PSD	c	r_f	r_A	f_1	α_1	α_2	μ	λ_0^+	D_{TB}/D_{DK}		
									$b=3$	$b=5$	$b=7$
Narrow band	1.05	n.a.	0	19.50	0.9999	0.9996	20	19.99	0.9999	0.9997	0.9996
	1.10	n.a.	0	19.01	0.9996	0.9985	20	19.97	0.999	0.999	0.998
Band limited	1.50	n.a.	0	15.50	0.993	0.975	20	19.50	0.991	0.981	0.969
	5	n.a.	0	5.14	0.933	0.827	20	16.54	0.980	0.920	0.867
	20	n.a.	0	1.29	0.886	0.765	20	15.29	0.998	0.932	0.875
	∞	n.a.	0	0.0	0.866	0.745	20	14.91	1.00	0.943	0.885
Bimodal	1.1/0.9	2.981	0.098	8.258	0.900	0.600	20	12.00	0.983	0.963	0.954
	1.1/0.9	7.408	0.006	4.677	0.900	0.300	20	6.00	0.929	0.966	0.964
	1.1/0.9	3.233	0.240	6.409	0.850	0.600	20	12.00	0.996	0.944	0.922
	1.1/0.9	7.176	0.050	2.920	0.700	0.300	20	6.00	1.000	0.989	0.992
	1.1/0.9	11.855	0.025	1.705	0.600	0.200	20	4.00	1.035	1.025	1.026
	1.1/0.9	11.715	0.051	1.628	0.550	0.250	20	5.00	1.060	1.029	1.027
	1.1/0.9	19.442	0.014	1.001	0.503	0.139	20	2.78	1.062	1.052	1.052

Among the examples of Table 1, two specific cases are investigated in more detail in Figures 6 and 7. They refer to combinations: $r_f = 7.408$, $r_A = 0.006$ (with $\alpha_1 = 0.900$, $\alpha_2 = 0.300$) and $r_f = 3.233$, $r_A = 0.240$ (with $\alpha_1 = 0.850$, $\alpha_2 = 0.600$). In Figures 6 and 7, the graphs on the left show the amplitude probability distributions $p_{DK}(s)$ and $p_{TB}(s)$, whereas the graphs on the right show each distribution multiplied by $\mu\, s^b/K$. In all figures, the Rayleigh distribution, also multiplied by $\left(\lambda_0^+\, s^b\right)/K$ on the right graph, is shown for comparison. Values $b = 5$ and $K = 1$ of the S/N curve are chosen.

The graphs on the right, so suitably devised, have the purpose of highlighting the distribution of the damage contributed by each amplitude. In fact, apart from constant K, the quantity $\mu\, s^b p(s)$ is the damage caused by the cycle with amplitude s. Looking at this kind of "probability distribution weighted with damage" or simply "damage distribution", it is possible to explain why, in spite of having different rainflow amplitude distributions, Dirlik and TB methods estimate almost identical damage values. For the example in Figure 6, the damage ratio is $D_{TB}/D_{DK} = 0.966$, while it slightly grows to $D_{TB}/D_{DK} = 0.944$ for the example in Figure 7. The comparison also demonstrates, at least for the two cases considered in the figures, that a difference in amplitude distributions in the region of small amplitudes is not of concern, since small amplitudes contribute very little to the total damage, especially for high b. Furthermore, the graph also makes apparent how the Rayleigh probability distribution always leads to a larger damage.

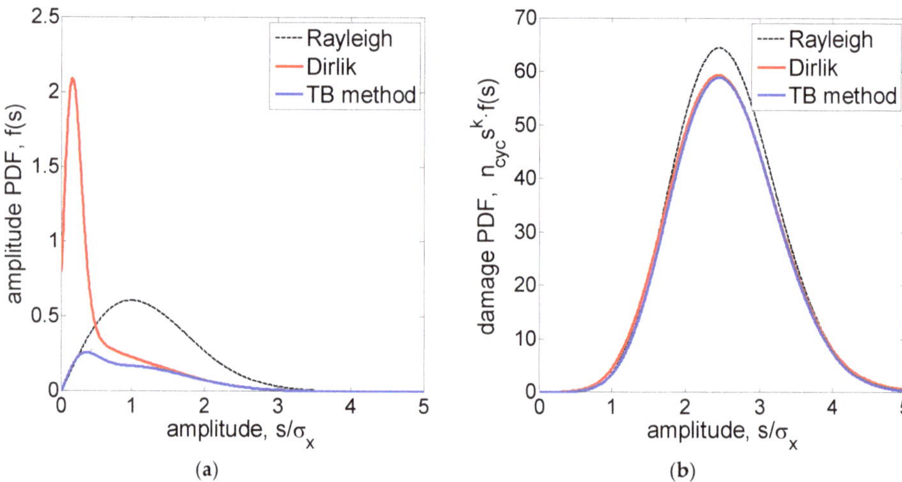

Figure 6. Comparison of (**a**) rainflow probability distributions, (**b**) weighted by the damage per amplitude. The figures refer to parameters $r_f = 7.408$, $r_A = 0.006$ (with $\alpha_1 = 0.900$, $\alpha_2 = 0.300$).

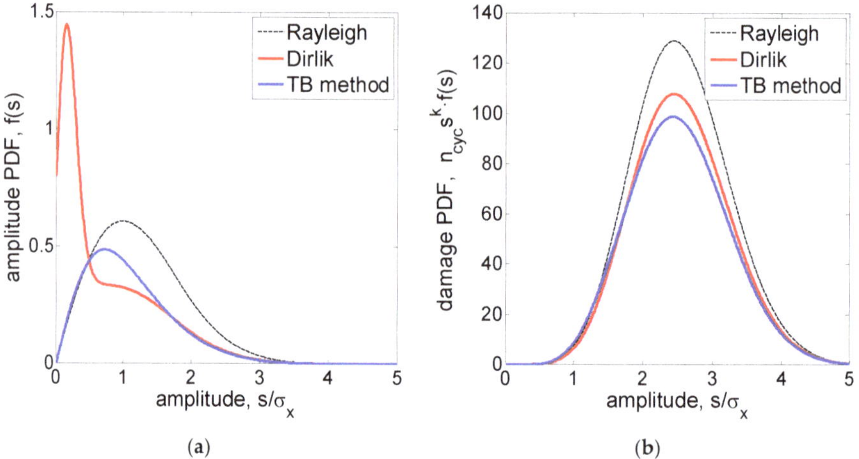

Figure 7. Comparison of (**a**) rainflow probability distributions, (**b**) weighted by the damage per amplitude. The figures refer to parameters $r_f = 3.233$, $r_A = 0.240$ (with $\alpha_1 = 0.850$, $\alpha_2 = 0.600$).

6.2. Comparison among Spectral Methods (Numerical Simulations)

To our knowledge, one of the earliest comparisons of spectral methods is that performed in [28]—obviously, it does not include the TB method that appeared only about ten years later. The comparison in [28] indicated the Dirlik method as superior to other spectral methods (as, for example, Wirsching-Light and "single-moment") for a wide combination of power spectral densities and S/N slopes, where its superiority was attributed to the use of four spectral moments. This outcome then highlighted the role of spectral moment m_1, in addition to m_0, m_2, and m_4, in describing the shape of the rainflow range probability distribution. To this regard, the article concluded that the rainflow distribution is a mixed distribution and, "in this sense Dirlik took the right approach" [28].

A later study [74], though actually not aiming to compare different methods, also confirmed the good performance of the Dirlik solution, even when applied to the subclass of bimodal random processes, for which specific solutions usually perform better.

The first comparison also including the TB method was presented in [17]. Using an error index to measure the estimation accuracy, the study confirmed that "the TB method has been shown to be as accurate as the DK approach "—although the latter has an error index slightly smaller. Actually, the lowest error is for TB method using w_3 in Equation (14), but surprisingly, this version has not become of common use. The results in [17] emphasize once more the need to include spectral moment m_1 to obtain improved damage estimates, although they have not yet excluded that a more complex relationship with other fractional moments, such as $m_{0.75}$ and $m_{1.5}$, could exist.

Of particular interest are the findings of [37], since this article compares spectral methods over a variety of power spectral density shapes that also include those already used in [15] and in [16,17] (see Figure 8). The results in this figure, quite similar to other figures in the article, show the ratio of the damage from spectral methods to the damage from time-domain simulations. In the figure, while the narrow band solution greatly overestimates the fatigue damage, Dirlik and TB methods both yield extremely accurate predictions, as in general occur also for other power spectra. Despite, as the article states, the fatigue damage predicted by the Dirlik and TB method "might be slightly underestimated in most of the cases" (the article here refers to some trimodal spectra), "it is demonstrated that the two formulae give quite accurate fatigue damage estimates for all of the spectral shapes and bandwidth parameters considered herein" [37].

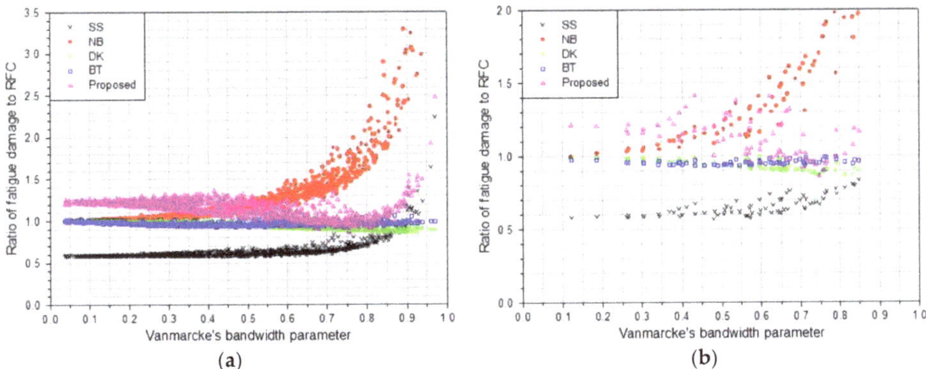

Figure 8. Example of comparison in [37] with the same power spectra used (**a**) in [16] and (**b**) in [15] (SS = simple sum of damage of frequency components in trimodal spectrum, NB = narrow band formula in Equation (5), DK = Dirlik, BT = TB method, Proposed = method proposed in [37]). Reproduced from [37] with permission from Elsevier.

The results in [33], too, corroborate the fact that both Dirlik and TB methods generally perform better than other approaches. The peculiarity of [33], however, is to have considered real power spectra as typical of structural dynamics and automotive industry. Unlike previous comparative studies, the findings of [33] seem to indicate a slightly better performance of TB method over Dirlik, although the difference remains within a few percentage points.

Common to previous studies is to observe for the tendency of prediction error to increase as the S/N curve becomes less steep (as b increases).

More recently, other simulation studies [77,78] have been carried out to benchmark the various spectral methods against different power spectral density shapes. Again, the comparison in [77] confirms that, for any spectral method, the estimation error is magnified with an increasing slope b, especially at values close to or above 8–10, which are typical of smooth/unnotched components. Contrasting trends characterize the accuracy of the Dirlik and TB methods, sometimes the former being better for some b values while the latter for other values.

In opposite trend with previous studies, the comparison of [78] claims different conclusions. Although for some power spectra the Dirlik and TB methods are "highlighted as the best-performing methods", in general their accuracy is judged not so satisfactory and, for steel and aluminum, it seems comparable to that of other methods—among them, is (surprisingly) also included the Wirsching-Light method, despite its accuracy with wide band spectra known to be not excellent.

6.3. Comparison with Experiments

Unlike simulation-based comparisons, there are far fewer studies that compare spectral methods against experiments. Only few of them are mentioned here.

An example is the experimental study in [79] focusing on bending-torsion random loading tests. Other than showing an agreement between estimations and experiments, this study also represents an example that explains how to apply spectral methods (Dirlik, specifically) in combination with multiaxial criteria based on an equivalent stress.

An interesting comprehensive experimental study is presented in [80], where estimations of several spectral methods were compared with experiments from bending and/or torsion random loadings with wide band power spectra of various shape. The multiaxial spectral criterion of [79] was first used to transform the local multiaxial stress into an equivalent uniaxial stress to which spectral methods (Dirlik, TB method, and others) are next applied to estimate the fatigue life by means of Smith-Watson-Topper (SWT) parameter [80]. The main conclusion of the experimental study is that the Dirlik and TB methods "proved to be substantially better than the Rayleigh method, and resulted in very small deviations when compared with the experimental data", as demonstrated by Figure 9.

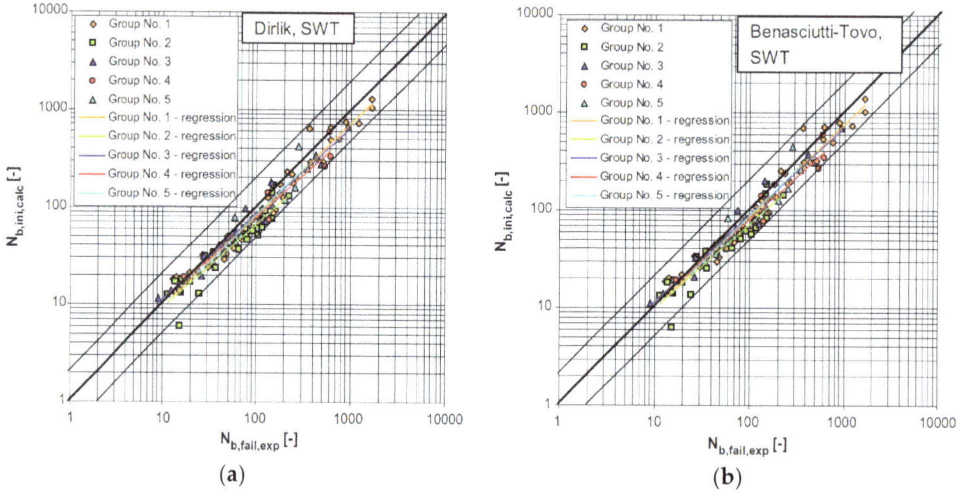

(a)   (b)

Figure 9. Example of comparison with experimental fatigue life presented in [80] for: (**a**) Dirlik method and (**b**) TB method. Reproduced from [80] with permission from Elsevier.

Interestingly, the study also pointed out how the two spectral methods "give a very similar fatigue life" [80], despite showing essential differences in their rainflow probability distributions. This apparent inconsistency is explained by the fact that both methods have in common that part of the rainflow probability distribution that contributes most to the fatigue damage—this aspect has already been discussed in Section 6.1.

A comparison with experimental data is provided in [61,62] for electronic devices subjected to random loadings, with particular attention on the structural integrity of solder joints in two architectures called package-on-package (PoP) and ball grid array (BGA). The comparison is displayed in Figure 10a. Although the number of test results is not so high to permit general conclusions to be drawn on a statistical basis, some trends emerge clearly. For example, in both the vibration tests the TB method yields estimations within 10% from experiments—note that the Dirlik method was not included in the study [61]. Similar trends were confirmed when other spectral methods were included in the comparison [62]. Again, the few experimental tests point out that, among all spectral methods, "both the Dirlik and Tovo–Benasciutti methods exhibit less than 10% absolute error, which are more accurate than others" [62]. For sure, these two methods are seen to perform much better than the Steinberg three-band method [73] that traditionally is preferred in the field of electronics applications.

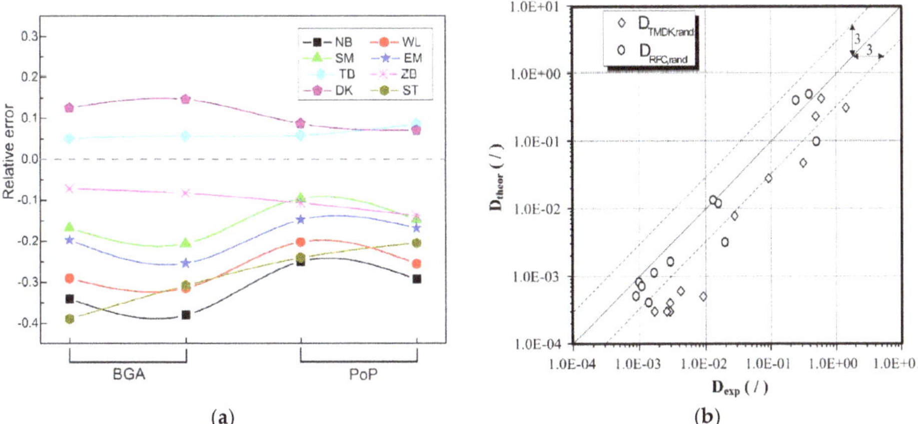

Figure 10. Comparison with experimental data: (**a**) electronic application for package-on-package (PoP) and ball grid array (BGA) (DK = Dirlik, TB = TB method, ZB = Zhao-Baker, SM = single-moment, ST = Steinberg method, WL = Wirsching-Light) (**b**) temperature effect on random loading (D_{TMDK} = temperature modified Dirlik method). Reproduced from [30,62] with permission from Elsevier.

As already mentioned in Section 3, the temperature-modified Dirlik method is checked against experimental data obtained in [30,31] by testing a die-casting aluminum alloy at various temperatures. Apart from some scatter in experimental data, the comparison shows that fatigue life can be estimated successfully by the Dirlik method [30,31] (see Figure 10b).

A nice correlation with experimental data, which confirms the capability of spectral methods in predicting fatigue life, is also highlighted in [81] for the Dirlik method—presumably the TB method, though not shown, is close to experimental data as well.

It finally has to be mentioned how the good accuracy of the Dirlik method encompasses not only metallic materials, but also advanced materials as composites [66].

7. Conclusions

Spectral methods are finding more and more application in various engineering areas and with various material types: from testing handheld devices to rocket launchers, from testing common metallic materials to ceramic matrix composites and even to concrete asphalt mortars.

Whether or not a spectral method is accurate enough in a certain application depends, first, on if its hypotheses are satisfied by that application. Methods restricted to narrow band processes will overestimate the damage if applied to a wide band process. Similarly, methods developed for stationary Gaussian processes will be wrong if applied to non-Gaussian or nonstationary processes. For any given application, the decision on which method is more appropriate is left to each one's engineering judgement.

On the other hand, a certain degree of approximation must be accepted when using spectral methods, since the actual measured random time histories of stress (for example, obtained by measurements) can hardly be exactly stationary and Gaussian, as instead prescribed by the models. The degree of approximation depends on how much the model hypotheses are violated. Sometimes, dividing a nonstationary signal into stationary, or almost-stationary, segments may be a simple and feasible solution. In other cases, especially with irregular time histories, the solution is not so obvious, or may even not exist.

Once it has been verified that a given application matches the model hypotheses, the choice of "the best" model sometimes becomes less obvious, especially for "newcomers" to vibration fatigue. The multitude of spectral methods in the literature may let one feel somehow disoriented, indeed. Help may come from results of comparative studies. As this article pointed out, for wide band stationary Gaussian processes, the literature seems to have found by now a consensus in identifying the methods by Dirlik and by Tovo–Benasciutti (TB) as "the best" spectral methods, with very little—if not even negligible—differences between them. Which one to use then becomes a matter of personal preference.

For what concerns possible areas of development of spectral methods, besides additional studies to evaluate the potential of spectral methods in different industrial sectors using different materials, some research topics could deserve attention in the future: multiaxial fatigue with nonstationary random loadings, crack propagation models, and, finally, the uncertainty in damage estimates due to sampling variability of power spectrum.

For multiaxial fatigue, it is known that the great majority of spectral criteria developed thus far—mostly as reformulations of multiaxial criteria originally in time domain [82,83]—are limited to stationary loadings. A great deal of research is required to extend those criteria to nonstationary multiaxial random loadings, by also considering that the nonstationary case has not yet been solved completely even for uniaxial loadings. At the same time, the development of new multiaxial spectral criteria, either in stationary or nonstationary case, should go hand in hand with experimental testing, aimed at gathering a database for calibrating and validating new multiaxial spectral methods—for example, by testing additively manufactured materials that have recently attracted so much attention by research.

For fatigue crack growth, it seems that the current literature on spectral-based crack propagation models is not so vast [6–10], so this topic may represent quite a promising research field; a starting point may be, for example, the development of spectral methods to include the effect of overloads.

Finally, a topic that, so far, has received very little or almost no attention is the effect on spectral fatigue damage of the sampling variability of a power spectral density computed from short time history records. When this sampling variability is included in the analysis, the fatigue damage estimated by spectral methods becomes a random value with a statistical variability evaluated through confidence intervals. Although some achievements have been obtained recently [84], this topic certainly deserves more attention by research in the years to come.

Author Contributions: Writing—original draft preparation, D.B.; writing—review and editing, T.D. and D.B. All authors have read and agreed to the published version of the manuscript.

Funding: This research received no external funding.

Conflicts of Interest: The authors declare no conflict of interest.

References

1. Matsuishi, M.; Endo, T. Fatigue of metals subjected to varying stress. Presented to Japan Society of Mechanical Engineers, Fukuoka, Japan, March 1968. In *The Rainflow Method in Fatigue: The Tatsuo Endo Memorial Volume*; Murakami, Y., Ed.; Butterworth-Heinemann Ltd.: Oxford, UK, 1992.
2. *ASTM E1049–85 (2017) Standard Practices for Cycle Counting in Fatigue Analysis*; ASTM International: West Conshohocken, PA, USA, 2017; Available online: www.astm.org (accessed on 23 August 2021).
3. Fatemi, A.; Yang, L. Cumulative fatigue damage and life prediction theories: A survey of the state of the art for homogeneous materials. *Int. J. Fatigue* **1998**, *20*, 9–34. [CrossRef]

4.	*MATLAB (2021), Version 9.10.0.1602886 (R2021a)*; The MathWorks Inc.: Natick, MA, USA, 2021; Available online: https://it.mathworks.com (accessed on 3 August 2021).
5.	Brodtkorb, P.A.; Johannesson, P.; Lindgren, G.; Rychlik, I.; Rydén, J.; Sjö, E. WAFO—A Matlab toolbox for analysis of random waves and loads. In Proceedings of the 10th Int. Offshore and Polar Engineering Conference (ISOPE), Seattle, WA, USA, 27 May–2 June 2000; Volume III, pp. 343–350.
6.	Zuccarello, B.; Adragna, N.F. A novel frequency domain method for predicting fatigue crack growth under wide band random loading. *Int. J. Fatigue* **2007**, *29*, 1065–1079. [CrossRef]
7.	Mao, W. Development of a spectral method and a statistical wave model for crack propagation prediction in ship structures. *J. Ship Res.* **2014**, *58*, 106–116. [CrossRef]
8.	Xue, X.; Chen, N.Z. Fracture mechanics analysis for a mooring system subjected to Gaussian load processes. *Eng. Struct.* **2018**, *162*, 188–197. [CrossRef]
9.	Zhang, Y.; Huang, X.; Wang, F. Fatigue crack propagation prediction for marine structures based on a spectral method. *Ocean Eng.* **2018**, *163*, 706–717. [CrossRef]
10.	Marques, D.E.T.; Vandepitte, D.; Tita, V. Damage detection and fatigue life estimation under random loads: A new structural health monitoring methodology in the frequency domain. *Fatigue Fract. Eng. Mater. Struct.* **2021**, *44*, 1622–1636. [CrossRef]
11.	Miles, J.W. On structural fatigue under random loading. *J. Aeron. Sci.* **1954**, *21*, 753–762. [CrossRef]
12.	Bendat, J.S. *Probability Functions for Random Responses: Prediction of Peaks, Fatigue Damage, and Catastrophic Failures*; Report NASA-CR-33; National Aeronautics and Space Administration (NASA): Washington, DC, USA, 1964.
13.	Benasciutti, D. *Fatigue Analysis of Random Loadings: A Frequency Domain Approach*; LAP Lambert Academic Publishing: Saarbrücken, Germany, 2012; ISBN 9783659123702.
14.	Slavič, J.; Boltezar, M.; Mrsnik, M.; Cesnik, M.; Javh, J. *Vibration Fatigue by Spectral Methods. From Structural Dynamics to Fatigue Damage—Theory and Experiments*; Elsevier: Amsterdam, The Netherlands, 2020.
15.	Dirlik, D. Application of Computers in Fatigue Analysis. Ph.D. Thesis, University of Warwick, Warwick, UK, 1985.
16.	Benasciutti, D.; Tovo, R. Spectral methods for lifetime prediction under wide-band stationary random processes. *Int. J. Fatigue* **2005**, *27*, 867–877. [CrossRef]
17.	Benasciutti, D.; Tovo, R. Comparison of spectral methods for fatigue analysis in broad-band Gaussian random processes. *Probab. Eng. Mech.* **2006**, *21*, 287–299. [CrossRef]
18.	Lutes, L.D.; Sarkani, S. *Random Vibrations: Analysis of Structural and Mechanical Systems*; Elsevier Butterworth-Heinemann: Oxford, UK, 2004.
19.	Troncossi, M.; Pesaresi, E. Analysis of synthesized non-Gaussian excitations for vibration-based fatigue life testing. *J. Phys. Conf. Ser.* **2019**, *1264*, 012039. [CrossRef]
20.	Vanmarcke, E.H. Properties of spectral moments with applications to random vibration. *J. Eng. Mech. Div.-ASCE* **1972**, *98(EM2)*, 425–446. [CrossRef]
21.	Hobbacher, A.F. *Recommendations for Fatigue Design of Welded Joints and Components*, 2nd ed.; IIW document IIW-1823-07 (ex XIII-2151r4-07/XV-1254r4-07); Springer International Publishing: Heidelberg, Germany, 2016.
22.	Wirsching, P.H.; Light, M.C. Fatigue under wide band random stresses. *J. Struct. Div. ASCE* **1980**, *106*, 1593–1607. [CrossRef]
23.	Google Scholar. Available online: https://scholar.google.com/scholar?hl=it&as_sdt=0%2C5&q=dirlik+fatigue&btnG= (accessed on 1 August 2021).
24.	Shinozuka, M. Monte Carlo solution of structural dynamics. *Comput. Struct.* **1972**, *2*, 855–874. [CrossRef]
25.	Shinozuka, M.; Deodatis, G. Simulation of stochastic processes by spectral representation. *Appl. Mech. Rev.* **1991**, *44*, 191–204. [CrossRef]
26.	Ibe, O.C. *Fundamentals of Applied Probability and Random Processes*, 2nd ed.; Academic Press: Cambridge, MA, USA, 2004.
27.	Sherratt, F.; Bishop, N.W.M.; Dirlik, T. Predicting fatigue life from frequency domain data. *Eng. Integr.* **2005**, *18*, 12–16.
28.	Bouyssy, V.; Naboishikov, S.M.; Rackwitz, R. Comparison of analytical counting methods for Gaussian processes. *Struct. Saf.* **1993**, *12*, 35–57. [CrossRef]
29.	Zalaznik, A.; Nagode, M. Frequency based fatigue analysis and temperature effect. *Mater. Des.* **2011**, *32*, 4794–4802. [CrossRef]
30.	Zalaznik, A.; Nagode, M. Validation of temperature modified Dirlik method. *Comput. Mater. Sci.* **2013**, *96*, 173–179. [CrossRef]
31.	Zalaznik, A.; Nagode, M. Experimental, theoretical and numerical fatigue damage estimation using a temperature modified Dirlik method. *Eng. Struct.* **2015**, *96*, 56–65. [CrossRef]
32.	Tovo, R. Cycle distribution and fatigue damage under broad-band random loading. *Int. J. Fatigue* **2002**, *24*, 1137–1147. [CrossRef]
33.	Mršnik, M.; Slavič, J.; Boltežar, M. Frequency-domain methods for a vibration-fatigue-life estimation—Application to real data. *Int. J. Fatigue* **2013**, *47*, 8–17. [CrossRef]
34.	Xu, J.; Zhang, Y.; Han, Q.; Li, J.; Lacidogna, G. Research on the scope of spectral width parameter of frequency domain methods in random fatigue. *Appl. Sci.* **2020**, *10*, 4715. [CrossRef]
35.	Benasciutti, D.; Tovo, R. Modelli di previsione del danneggiamento a fatica per veicoli in regime stazionario ed ergodico. In Proceedings of the Italian Association for Stress Analysis (AIAS), Parma, Italy, 18–20 September 2002. (In Italian).
36.	Benasciutti, D.; Tovo, R. Spectral methods for lifetime prediction under wide-band stationary random processes. In Proceedings of the Int. Conference "Cumulative Fatigue Damage", Seville, Spain, 27–29 May 2003; p. 46.

37. Gao, Z.; Moan, T. Frequency-domain fatigue analysis of wide-band stationary Gaussian processes using a trimodal spectral formulation. *Int. J. Fatigue* **2008**, *30*, 1944–1955. [CrossRef]

38. Zhao, W.; Baker, M.J. On the probability density function of rainflow stress range for stationary Gaussian processes. *Int. J. Fatigue* **1992**, *14*, 121–135. [CrossRef]

39. Lutes, L.D.; Larsen, C.E. Improved spectral method for variable amplitude fatigue prediction. *J. Struct. Eng.-ASCE* **1990**, *116*, 1149–1164. [CrossRef]

40. Larsen, C.E.; Lutes, L.D. Predicting the fatigue life of offshore structures by the single-moment method. *Probab. Eng. Mech.* **1991**, *6*, 96–108. [CrossRef]

41. Yaich, A.; El Hami, A. Numerical and experimental investigation on multiaxial fatigue damage estimation of Qualmark chamber test table structures under random vibrations. *Mech. Based Des. Struct. Mech.* **2021**. published online. [CrossRef]

42. Niesłony, A.; Macha, E. *Spectral Method in Multiaxial Random Fatigue*; Springer: Berlin, Germany, 2007.

43. Bel Knani, K.; Benasciutti, D.; Signorini, A.; Tovo, R. Fatigue damage assessment of a car body-in-white using a frequency-domain approach. *Int. J. Mater. Prod. Technol.* **2007**, *30*, 172–198. [CrossRef]

44. Valsamos, G.; Theodosiou, C.; Natsiavas, S. Periodic steady state response and fatigue analysis of a nonlinear city bus model. In Proceedings of the ASME 2009 International Design Engineering Technical Conferences and Computers and Information in Engineering Conference, San Diego, CA, USA, 30 August–2 September 2009. [CrossRef]

45. Braccesi, C.; Cianetti, F.; Lori, G.; Pioli, D. Random multiaxial fatigue: A comparative analysis among selected frequency and time domain fatigue evaluation methods. *Int. J. Fatigue* **2015**, *74*, 107–118. [CrossRef]

46. Braccesi, C.; Cianetti, F.; Tomassini, L. An innovative modal approach for frequency domain stress recovery and fatigue damage evaluation. *Int. J. Fatigue* **2016**, *91*, 382–396. [CrossRef]

47. Nascimento, V.; Teixeira, G.; Clarke, T. Structural validation of a pneumatic brake actuator using method for fatigue life calculation. *Eng. Fail. Anal.* **2020**, *118*, 104837. [CrossRef]

48. Kong, Y.S.; Abdullah, S.; Schramm, D.; Omar, M.Z.; Haris, S.M. Vibration fatigue analysis of carbon steel coil spring under various road excitations. *Metals* **2018**, *8*, 617. [CrossRef]

49. Wang, N.; Liu, J.; Zhang, Q.; Yang, H.; Tang, M. Fatigue life evaluation and failure analysis of light beam direction adjusting mechanism of an automobile headlight exposed to random loading. *Proc. Inst. Mech. Eng. Part D-J. Automob. Eng.* **2019**, *233*, 224–231. [CrossRef]

50. Cima, M.; Solazzi, L. Experimental and analytical study of random fatigue, in time and frequencies domain, on an industrial wheel. *Eng. Fail. Anal.* **2021**, *120*, 105029. [CrossRef]

51. Ugras, R.C.; Alkan, O.K.; Serkan Orhan, S.; Kutlu, M.; Mugan, A. Real time high cycle fatigue estimation algorithm and load history monitoring for vehicles by the use of frequency domain methods. *Mech. Syst. Signal Proc.* **2019**, *118*, 290–304. [CrossRef]

52. DNV GL. *Class Guideline DNVGL-CG-0129, Fatigue Assessment of Ship Structures*; DNV GL: Oslo, Norway, 2015; Available online: www.dnv.com (accessed on 23 August 2021).

53. Bittar Filho, J.H.; Lourenço de Souza, M.I.; Pasqualino, I.P. Comparison of spectral fatigues methodologies using equivalent stresses obtained by Mises and Battelle applied to a semi-submersible platform. In *Practical Design of Ships and Other Floating Structures*; Okada, T., Suzuki, K., Kawamura, Y., Eds.; Springer: Singapore, 2019; Volume 64, pp. 580–607. [CrossRef]

54. Lynch, J.P.; Law, K.H.; O'Connor, S. *Probabilistic & Reliability-Based Health Monitoring Strategies for High-Speed Naval Vessels*; Report N00014-09-1-0567; Office of Naval Research (ONR): Arlington, VA, USA, 2012.

55. Yeter, B.; Garbatov, Y.; Guedes Soares, C. Evaluation of fatigue damage model predictions for fixed offshore wind turbine support structures. *Int. J. Fatigue* **2016**, *87*, 71–80. [CrossRef]

56. Böhm, M.; Kowalski, M. Fatigue life estimation of explosive cladded transition joints with the use of the spectral method for the case of a random sea state. *Mar. Struct.* **2020**, *71*, 102739. [CrossRef]

57. Ragan, P.; Lance, M. Comparing estimates of wind turbine fatigue loads using time-domain and spectral methods. *Wind Eng.* **2007**, *31*, 83–99. [CrossRef]

58. Chen, X. Analysis of crosswind fatigue of wind-excited structures with nonlinear aerodynamic damping. *Eng. Struct.* **2014**, *74*, 145–156. [CrossRef]

59. Huo, T.; Tong, L. An approach to wind-induced fatigue analysis of wind turbine tubular towers. *J. Constr. Steel. Res.* **2020**, *166*, 105917. [CrossRef]

60. Fekih, L.B.; Kouroussis, G.; Verlinden, O. Spectral-based fatigue assessment of ball grid arrays under aerospace vibratory environment. *Key Eng. Mater.* **2013**, *569–570*, 425–432. [CrossRef]

61. Xia, J.; Li, G.; Li, B.; Cheng, L.; Zhou, B. Fatigue life prediction of Package-on-Package stacking assembly under random vibration loading. *Microelectron. Reliab.* **2017**, *71*, 111–118. [CrossRef]

62. Xia, J.; Yang, L.; Liu, Q.; Peng, Q.; Cheng, L.; Li, G. Comparison of fatigue life prediction methods for solder joints under random vibration loading. *Microelectron. Reliab.* **2019**, *95*, 58–64. [CrossRef]

63. Gao, D.Y.; Yao, W.X.; Wen, W.D.; Huang, J. Equivalent spectral method to estimate the fatigue life of composite laminates under random vibration loadings. *Mech. Compos. Mater.* **2021**, *57*, 101–114. [CrossRef]

64. Chen, X.; Sun, Y.; Wu, Z.; Yao, L.; Zhang, Y.; Zhou, S.; Liu, Y. An investigation on residual strength and failure probability prediction for plain weave composite under random fatigue loading. *Int. J. Fatigue* **2019**, *120*, 267–282. [CrossRef]

65. Sun, Y.; Zhang, Y.; Yang, C.; Liu, Y.; Chen, X.; Yao, L.; Gao, W. Prediction on fatigue properties of the plain weave composite under broadband random loading. *Fatigue Fract. Eng. Mater. Struct.* **2021**, *44*, 1515–1532. [CrossRef]
66. Böhm, M.; Głowacka, K. Fatigue life estimation with mean stress effect compensation for lightweight structures—The case of GLARE 2 composite. *Polymers* **2020**, *12*, 251. [CrossRef]
67. Kim, H.; Kim, G.; Ji, W.; Lee, Y.S.; Jang, S.; Shin, C.M. Random vibration fatigue analysis of a multi-material battery pack structure for an electric vehicle. *Funct. Compos. Struct.* **2021**, *3*, 025006. [CrossRef]
68. Zhou, Y.; Wu, S.; Tan, Z.; Fei, Q. Temperature-dependence of acoustic fatigue life for thermal protection structures. *Theor. Appl. Mech. Letters* **2014**, *4*, 021005. [CrossRef]
69. Česnik, M.; Slavič, J.; Boltežar, M. Assessment of the fatigue parameters from random vibration testing: Application to a rivet joint. *Strojniski Vestn.-J. Mech. Eng.* **2016**, *62*, 471–482. [CrossRef]
70. Go, E.-S.; Kim, M.-G.; Kim, I.-G.; Kim, M.-S. Fatigue life prediction in frequency domain using thermal-acoustic loading test results of titanium specimen. *J. Mech. Sci. Technol.* **2020**, *34*, 4015–4024. [CrossRef]
71. Ding, H.; Zhu, Q.; Zhang, P. Fatigue damage assessment for concrete structures using a frequency-domain method. *Math. Probl. Eng.* **2014**, *2014*, 407193. [CrossRef]
72. Ortiz, K.; Chen, N.K. Fatigue damage prediction for stationary wide-band stresses. In *Reliability and Risk Analysis in Civil Engineering, Proceedings of the 5th International Conference on the Applications of Statistics and Probability in Soil and Structural Engineering (ICASP5), Vancouver, Canada, 25–29 May 1987*; Institute for Risk Research, University of Waterloo: Waterloo, ON, Canada, 1987.
73. Steinberg, D.S. *Vibration Analysis for Electronic Equipment*, 3rd ed.; Wiley-Interscience: New York, NY, USA, 2000.
74. Fu, T.-T.; Cebon, D. Predicting fatigue lives for bi-modal stress spectral densities. *Int. J. Fatigue* **2000**, *22*, 11–21. [CrossRef]
75. Benasciutti, D.; Tovo, R. On fatigue damage assessment in bimodal random processes. *Int. J. Fatigue* **2007**, *29*, 232–244. [CrossRef]
76. Lalanne, C. *Mechanical Vibration and Shock*; Hermes Penton Science: London, UK, 2009.
77. Larsen, C.E.; Irvine, T. A review of spectral methods for variable amplitude fatigue prediction and new results. *Procedia Eng.* **2015**, *101*, 243–250. [CrossRef]
78. Han, Q.; Li, J.; Xu, J.; Ye, F.; Carpinteri, A.; Lacidogna, G. A new frequency domain method for random fatigue life estimation in a wide-band stationary Gaussian random process. *Fatigue Fract. Eng. Mater. Struct.* **2019**, *42*, 97–113. [CrossRef]
79. Niesłony, A. Comparison of some selected multiaxial fatigue failure criteria dedicated for spectral method. *J. Theor. Appl. Mech.* **2010**, *48*, 233–254.
80. Niesłony, A.; Růžička, M.; Papuga, J.; Hodr, A.; Balda, M.; Svoboda, J. Fatigue life prediction for broad-band multiaxial loading with various PSD curve shapes. *Int. J. Fatigue* **2012**, *44*, 74–88. [CrossRef]
81. Gadolina, I.V.; Makhutov, N.A.; Erpalov, A.V. Varied approaches to loading assessment in fatigue studies. *Int. J. Fatigue* **2021**, *144*, 106035. [CrossRef]
82. Benasciutti, D.; Sherratt, F.; Cristofori, A. Recent developments in frequency domain multi-axial fatigue analysis. *Int. J. Fatigue* **2016**, *91*, 397–413. [CrossRef]
83. Carpinteri, A.; Spagnoli, A.; Vantadori, S. A review of multiaxial fatigue criteria for random variable amplitude loads. *Fatigue Fract. Eng. Mater. Struct.* **2017**, *40*, 1007–1036. [CrossRef]
84. Benasciutti, D. Confidence interval of the 'single-moment' fatigue damage calculated from an estimated power spectral density. *Int. J. Fatigue* **2021**, *145*, 106131. [CrossRef]

*metals*

Article

Fatigue Analysis of Nonstationary Random Loadings Measured in an Industrial Vehicle Wheel: Uncertainty of Fatigue Damage

Julian M. E. Marques [1,*], **Luigi Solazzi** [2] and **Denis Benasciutti** [3,*]

[1] Department of Mechanics, Biomechanics and Mechatronics, Faculty of Mechanical Engineering, Czech Technical University, Technická 4, 166 36 Prague, Czech Republic

[2] Department of Industrial and Mechanical Engineering, University of Brescia, Via Branze 38, 25123 Brescia, Italy; luigi.solazzi@unibs.it

[3] Department of Engineering, University of Ferrara, Via Saragat 1, 44122 Ferrara, Italy

* Correspondence: julianmarcell.enzveilermarques@fs.cvut.cz (J.M.E.M.); denis.benasciutti@unife.it (D.B.); Tel.: +39-0532-974976 (D.B.)

Abstract: This article presents an application of a method for estimating the inherent statistical variability of the fatigue damage computed in one single nonstationary random time history. The method applies the concept of confidence interval for the damage, which is constructed after the single time history is subdivided into pseudo-stationary segments, with each of them further divided into shorter blocks. As a case study, the method is applied to the strain time histories measured in a wheel of a telescopic handler industrial vehicle. A preliminary screening involving the short-time Fourier transform and the run test is carried out to verify whether the measured time histories are truly nonstationary and fall within the hypotheses of the proposed method. After that, the confidence interval for the unknown expected damage is computed; its upper bound can be used as a safety limit in a structural integrity assessment. The obtained results seem very promising and suggest the possible use of the proposed approach in similar engineering applications.

Keywords: fatigue damage; nonstationary random loadings; run test; short-time Fourier transform

Citation: Marques, J.M.E.; Solazzi, L.; Benasciutti, D. Fatigue Analysis of Nonstationary Random Loadings Measured in an Industrial Vehicle Wheel: Uncertainty of Fatigue Damage. *Metals* **2022**, *12*, 616. https://doi.org/10.3390/met12040616

Academic Editors: Mark T. Whittaker, Turan Dirlik and Anders Jarfors

Received: 25 February 2022
Accepted: 30 March 2022
Published: 2 April 2022

Publisher's Note: MDPI stays neutral with regard to jurisdictional claims in published maps and institutional affiliations.

1. Introduction

Wheels in industrial vehicles are critical components, being subjected in service to complex random and nonstationary fatigue loadings. Here, the term 'random' indicates the intrinsic aleatory nature of the loading that makes it impossible to exactly predict the loading values. Instead, 'nonstationary' specifies that the statistical properties of the random loading (e.g., mean value, root-mean-square value, autocorrelation function) change over time as a consequence, for example, of the different driving conditions, routes (curves, straights, jolts, etc.), or vehicle usage (e.g., speed, operating modes) [1–4].

When considering the design of industrial vehicles from a material viewpoint, a new trend has recently emerged that exploits a lightweight design based on the use of composite materials not only for wheels [5,6] but also for other structural details of the vehicle (e.g., trucks [7,8], earthmoving machines [9], or working platforms [10,11]). This trend in design aims at improving the wheel and vehicle performance, while reducing the overall weight.

In addition to the material choice, a crucial step in the design process remains the structural integrity assessment of critical parts. For wheels of industrial vehicles, the design process usually develops as a synergistic interaction between numerical modeling and in-field experimental testing [12,13]. The outcomes of numerical simulation analyses are, indeed, supported by experimental tests conducted on prototypes or real vehicles subjected to standardized loadings that should be as representative as possible of the actual loadings observed in service.

Within this experimental framework, one of the most complex tasks is the definition of the testing conditions that lead to the most representative loadings. An example of the complexity of service loadings is that characterizing the wheels of industrial vehicles in industrial sectors such as agricultural, construction, or material-handling. In these sectors, the wheels have to endure heavy loads characterized by different intensities or that are distributed unevenly as a result of the excitation from an uneven ground. The complexity in loading definition further increases if the loading intensity is correlated to the various wheel geometries that often cover a wide range; for example, the nominal wheel load can range from 4 kN up to 250 kN for a wheel size varying from $8''$ (203 mm) to $54''$ (1371 mm) in diameter and from $3''$ to $36''$ in width [12].

Accordingly, it becomes quite unlikely for a single validation test to be able to encompass all the loading conditions to which the wheels will be subjected in their service life, or to include all the number of cycles counted in the entire wheel lifetime; this number is on the order of several millions as determined by the number of wheel revolutions.

If interpreted from a statistical point of view, the definition of appropriate testing loadings for experimental validation becomes a matter of the representativeness and sampling variability of the testing loading. When only a single or, at best, a few loading time histories are recorded from tests, reliable conclusions on the service fatigue life of the wheel can only be made on a statistical basis from the knowledge of a small set of fatigue cycles and few damage values. Such few damage values, however, represent only a small fraction of a much larger population of values that might theoretically be obtained from testing a much larger (virtually infinite) set of time histories—a situation clearly not obtainable in practice. The goal is to deduce meaningful statistical conclusions on the wheel reliability and safety solely from the knowledge of such a small sample of time history records and damage values obtained in experiments.

This issue is intimately related to the inherent statistical variability of fatigue damage. More precisely, let $D(T)$ be the fatigue damage of a time history $z(t)$ of time duration T. From a mathematical point of view, damage $D(T)$ is a random variable that follows a certain probability distribution with expected value $E[D(T)]$ and variance σ_D^2.

Starting from the pioneering works of Bendat [14] and Mark and Crandall [15,16], several methods have been proposed for estimating $E[D(T)]$ and σ_D^2 from the statistical properties of random loading in the time domain or from its power spectral density in the frequency domain. While the early works were restricted to certain classes of random processes (Gaussian linear oscillator) [14–16], more recent methods made the solution include any kind of Gaussian random loading with specific types of power spectral density, e.g., narrow-band, multimodal, or wide-band [17–19]. An extension to narrow-band and non-Gaussian cases has also been proposed recently [20].

A completely different approach, developed in [21], exploited a data-driven analysis of time history records to account for the variability of the random loading by constructing a confidence interval for the expected damage. The approach, first benchmarked with numerical simulations, was further validated against experimental records for a mountain bike under pseudo-stationary conditions [22].

The approaches mentioned so far are, however, restricted to the case of stationary random loadings with time-invariant statistical properties. This stationary condition, as already emphasized, may be too oversimplified for the nonstationary in-service loadings experienced by the wheels of industrial vehicles.

This work, thus, attempts to extend the confidence interval approach mentioned above to the case of nonstationary random loadings. Although the theoretical framework of the method was preliminary devised in [23,24], the goal here is to apply the method to an actual case study, i.e., to the nonstationary time histories recorded from a wheel of a telescopic handler vehicle.

In a preliminary stage, each recorded time history is confirmed to be really nonstationary by the application of, first, the short-time Fourier transform (STFT) and, then, the run test. The outcome of the STFT allows the time history to be divided into stationary or pseudo-stationary segments. On the basis of such subdivision, it is possible to apply, for each state, the usual method valid for stationary loading to compute the confidence interval for the expected damage. Confidence intervals for different stationary segments are combined to arrive at the confidence interval for the whole nonstationary time history. The obtained results seem very promising and suggest the possible use of the proposed approach in similar engineering applications.

2. Fatigue Damage and Expected Value

Let $z(t)$, $0 \leq t \leq T$ be a nonstationary time history of time duration T, which acts on a component or structural detail. According to the Palmgren–Miner damage accumulation rule, the fatigue damage of $z(t)$ is computed by summing up the damage of all rainflow cycles

$$D(T) = \sum_{i=1}^{N_T} d_i = \sum_{i=1}^{N_T} \frac{S_i^m}{K} \tag{1}$$

where S_i is the stress amplitude of the i-th rainflow cycle, and N_T is the total number of rainflow cycles counted in $z(t)$. Parameter K is the fatigue strength coefficient, and m is the inverse slope of the S–N curve $S^m N = K$ of the component or structural detail. Below, a single-slope S–N curve is used, but the approach presented hereafter allows for any type of S–N curve. Furthermore, to account for the scatter in material fatigue strength, the S–N curve $S^m N = K$ can be thought to represent a characteristic line defined for a prescribed survival probability and confidence [25].

Since $z(t)$ is a random time history, the fatigue cycles counted in $z(t)$ have randomly distributed amplitudes, and their number N_T is also random. As a result, the damage $D(T)$ in Equation (1) has to be regarded as a random variable, being affected by two sources of randomness related to amplitudes S_i and number of cycles N_T. In other words, the random variables, S and N_T, determine an inherent statistical variability of the damage value $D(T)$ when it is computed from a particular time history $z(t)$. For example, damage $D(T)$ is likely not to be equal if it were computed from other time histories, despite being obtained or measured under virtually identical conditions. From a probabilistic point of view, the random variable $D(T)$ follows a certain probability distribution with expected value $E[D(T)]$ and variance σ_D^2.

As a result, the damage $D(T)$ computed from a single time history $z(t)$ with finite length T is only a sample value that is distributed randomly around $E[D(T)]$; it may be smaller or larger than $E[D(T)]$. If the ratio $\sigma_D / \sqrt{E[D(T)]}$ (coefficient of variation) is large, the values of $D(T)$ may be so scattered around $E[D(T)]$ that a fatigue life estimation based on a single value $D(T)$ becomes rather uncertain [21].

Since the damage $D(T)$ is a random variable following a certain distribution, the expected damage is given by taking the expectation of Equation (1)

$$E[D(T)] = E\left[\sum_{i=1}^{N_T} d_i\right] = E[N_T]\frac{E[S^m]}{K}, \tag{2}$$

where $E[\cdot]$ is the probabilistic expectation, which represents a weighted average over an infinite population of values.

According to this definition of expectation, the expected damage $E[D(T)]$ in Equation (2) is the average over an infinite population of damage values that are computed from an infinite ensemble of time histories; it clearly represents a mathematical abstraction. In engineering applications, conversely, the common situation is that in which only one time history with duration T is obtained from measurements. In this case, $E[D(T)]$ cannot be computed since the infinite ensemble of time histories is unknown.

The problem then becomes that to make statistical conclusions about $E[D(T)]$ solely on the basis of the knowledge of damage $D(T)$. Since this damage is a random variable with expected value $E[D(T)]$, it is not known a priori the closeness of $D(T)$ and $E[D(T)]$. The damage $D(T)$ can be very close to $E[D(T)]$, or it can be significantly far from $E[D(T)]$. A way to address this issue is to use $D(T)$ for constructing a confidence interval for $E[D(T)]$.

In [21], a $100(1 - \beta)\%$ confidence interval to include $E[D(T)]$ was proposed for the case of a stationary time history. The proposed confidence interval was proven to agree with results from simulated [21] and measured stationary loadings, the latter acquired on a mountain bike [22]. Despite these promising results, the proposed approach was only applied to stationary time histories. In situations where the time history is nonstationary, a different procedure is needed to obtain the confidence interval for $E[D(T)]$.

3. Confidence Interval of Damage in a Nonstationary Loading

This section describes how to construct a confidence interval to enclose the expected damage for the case of a nonstationary switching random time history $z(t)$. The term 'switching' indicates that $z(t)$ is formed by a sequence of stationary load states. The number of load states needs to be two or more, $N_S \geq 2$, and they can have same or different time durations, $T_{S,i}, i = 1, 2, \ldots N_S$. The total time length of $z(t)$ is $T = \sum_{i=1}^{N_S} T_i$.

Note that the different states can appear in any random sequence in $z(t)$ and any number of times (see Figure 1a), i.e., the same state can appear repeatedly several times in the sequence, as often observed in real applications [26]. In this circumstance, the quantity $T_{S,i}$ indicates the *total* time duration of state i in $z(t)$.

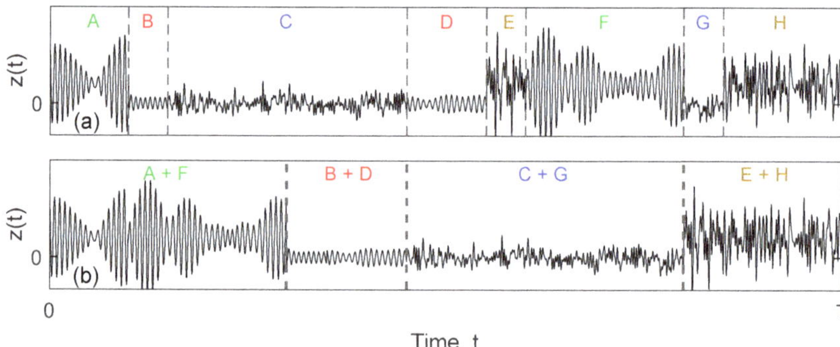

Figure 1. Example of state reordering: (**a**) nonstationary switching time history with replicated stationary states; (**b**) the same time history after state reordering.

For the procedure that follows, it is irrelevant in which time order and how many times a stationary state appears in $z(t)$. Hence, in a very first stage, the original sequence of states must be reordered so that, in the sequence, each load state appears only once in its full length $T_{S,i}$. This stage reordering yields a nonstationary time history $z(t)$ without replicated load states (see the example in Figure 1b). While this state reordering will greatly simplify the following analysis, it only causes a small number of cycles to be lost, i.e., those cycles formed by peaks and valleys falling in different states before reordering.

After reordering, the nonstationary time history with load states ordered in sequence is divided into N_S disjoint states of length $T_{S,i}, i = 1, 2, \ldots N_S$. A further subdivision of each load state into N_B blocks is then performed; the blocks must be disjointed (not overlapped). Symbol $T_{B,i}$ denotes the time length of bocks in state i. Note that, since each state has in principle a different time length, the use of the same number of block subdivisions for all states makes the blocks in each state have different time lengths.

Within the i-th state, the damage of each block is computed by the rainflow counting method and Palmgren–Miner rule. After subdivision into states and blocks, $N_S \cdot N_B$ damage values $D_{B,ij}(T_{B,i}), i = 1, 2, \ldots N_S$ and $j = 1, 2, \ldots N_B$ are obtained, where index i identifies the state and j denotes the block.

Since the states are distinct and the blocks are not overlapped, the damage values $D_{B,ij}(T_{B,i})$ are independent random variables. Furthermore, since each state is stationary, the damage values $D_{B,ij}(T_{B,i})$, $j = 1, 2, \ldots N_B$ of the blocks within the same i-th stationary state follow the same probability distribution, i.e., they are identically distributed.

A further hypothesis, also adopted in [18,21], is that the block damage values in each state follow a normal probability distribution, which may be different for each stationary state. For the block damage values $D_{B,ij}(T_{B,i})$, the sample mean $\overline{D}_{B,i}(T_{B,i})$ and the sample variance $\hat{\sigma}^2_{D_B,i}$ are computed as

$$\overline{D}_{B,i}(T_B) = \frac{1}{N_B} \sum_{j=1}^{N_B} D_{B,ij}(T_{B,i}),$$

$$\hat{\sigma}^2_{D_B,i} = \frac{1}{N_B-1} \sum_{j=1}^{N_B} \left[D_{B,ij}(T_{B,i}) - \overline{D}_{B,i}(T_{B,i}) \right]^2. \tag{3}$$

The technique of dividing the whole nonstationary time history into stationary states and each state into blocks, and then calculating, for each state, the corresponding sample values of the damage represents a preliminary data processing stage required for the confidence interval of damage to be constructed.

This confidence interval is obtained by reformulating the approximate confidence interval for the difference in means of two independent normal random variables with unknown and unequal variances [27]. Instead of the difference of the means of two variables, the confidence interval here proposed considers the sum in means of two or more random variables. Fortunately, the difference and the sum of normal random variables are also normally distributed.

According to this reformulation, the confidence interval of the expected damage is as follows [23,24]:

$$\sum_{i=1}^{N_S} \overline{D}_{B,i}(T_{B,i}) - t_{\beta/2,\nu} \sqrt{\sum_{i=1}^{N_S} \frac{\hat{\sigma}^2_{D_B,i}}{N_B}} \leq \sum_{i=1}^{N_S} E[D_{B,i}(T_{B,i})]$$

$$\leq \sum_{i=1}^{N_S} \overline{D}_{B,i}(T_{B,i}) + t_{\beta/2,\nu} \sqrt{\sum_{i=1}^{N_S} \frac{\hat{\sigma}^2_{D_B,i}}{N_B}}, \tag{4}$$

in which $t_{\beta/2,\nu}$ is the quantile of Student's t-distribution, and ν is an equivalent number of degrees of freedom

$$\nu \cong (N_B - 1) \frac{\left(\sum_{i=1}^{N_S} \hat{\sigma}^2_{D_B,i} \right)^2}{\sum_{i=1}^{N_S} \left(\hat{\sigma}^2_{D_B,i} \right)^2}. \tag{5}$$

After substituting the sample mean $\overline{D}_{B,i}(T_{B,i})$ into Equation (4) and multiplying this expression by N_B, the expression of the confidence interval turns out to be

$$\sum_{i=1}^{N_S} \sum_{j=1}^{N_B} D_{B,ij}(T_{B,i}) - t_{\beta/2,\nu} \sum_{i=1}^{N_S} \sqrt{N_B \cdot \hat{\sigma}^2_{D_B,i}} \leq \sum_{i=1}^{N_S} N_B E[D_{B,i}(T_{B,i})]$$

$$\leq \sum_{i=1}^{N_S} \sum_{j=1}^{N_B} D_{B,ij}(T_{B,i}) + t_{\beta/2,\nu} \sum_{i=1}^{N_S} \sqrt{N_B \cdot \hat{\sigma}^2_{D_B,i}}. \tag{6}$$

Since N_S and N_B can be considered as deterministic values (not random), it is possible to substitute into Equation (6) the single summation of the expected damage of blocks $\sum_{i=1}^{N_S} N_B E[D_{B,i}(T_{B,i})] = E[D(T)]$ and the double summation $\sum_{i=1}^{N_S} \sum_{j=1}^{N_B} D_{B,ij}(T_{B,i}) \cong D(T)$. The double summation well approximates the total damage $D(T)$ since only a small number of cycles are lost after subdivision of the time history into blocks. To ensure that this approximation is acceptable, the number of cycles in each block should be much greater than the number of blocks (see [21]), a condition that is usually met in practice.

After the above mathematical simplifications, the final expression of the $100(1 - \beta)\%$ confidence interval of the expected damage $E[D(T)]$ for a single nonstationary switching time history is

$$D(T) - t_{\beta/2,\nu} \sum_{i=1}^{N_S} \sqrt{N_B \cdot \hat{\sigma}^2_{D_B,i}} \leq E[D(T)] \leq D(T) + t_{\beta/2,\nu} \sum_{i=1}^{N_S} \sqrt{N_B \cdot \hat{\sigma}^2_{D_B,i}}. \tag{7}$$

This confidence interval can be constructed if a minimum number of blocks $N_B \geq 2$ is available to compute the sample statistics. For $N_S = 1$, Equation (7) yields the result obtained in [21].

In the next sections, this confidence interval is applied to a nonstationary time history measured in a wheel of a telescopic handler vehicle.

4. Experimental Tests

The experimental tests were carried out on the frontal wheel of the loader Manitou MT 1840 (Manitou BF, Ancenis, France) (see Figure 2). The main characteristics of the machine are as follows: weight in unladen conditions 116.3 kN; working load limit on the forks 40 kN; weight in the front axle equal to 54.8 kN in unloaded condition and 129.3 kN with the maximum load on the forks.

Figure 2. Different load conditions applied to the machine: (**a**) on-road case; (**b**) load placement case; (**c**) off-road case; (**d**) strain gauges and data system applied to the wheel to acquire and register the strain time-histories.

The wheel positioned on the machine is a classical one made of S355 structural steel (UNI EN 10025). The wheel diameter is about 1020 mm, and the width is about 320 mm. On the wheel, a Michelin Power Cl 440/80–24" 168 A8 tire (Michelin S.A., Clermont Ferrand, France) was mounted, with an inflate pressure of 4.5 bar. Other details can be found in [12].

As already pointed out, there are many different load conditions that a machine can be subjected to during its life and they depend on several factors such as the route type, velocity, and weight. In the absence of specific standards that help define the load conditions to use for designing the machine and its structural components, four different load cases were analyzed in the experiments. These load conditions were chosen, in agreement with the wheel manufacturer, with the aim to better describe the different uses to which the machine can be subjected during its life.

1. On-road case: the machine moved on a paved road in an oval circuit; this load condition was performed without any load on the forks (Figure 2a);
2. Load placement case: this condition is similar to the previous one except that 20 kN loads at different heights were applied on the loader (Figure 2b);
3. Loader case: in this case, only 40 kN was applied to the machine's forks at a fixed height;
4. Off-road case: the machine moved on an irregular road, with curves and slaloms. The load on the fork was equal to 15 kN, Figure 2c.

Strain gauges (Vishay Precision Group Inc., Raleigh, NC, USA) and an appropriate data logger (Hottinger Bruel & Kjaer, Virum, Denmark) were applied on the wheel to acquire and record the strain signals (see Figure 2d). The strain gauges applied to the wheel were both uniaxial and biaxial (see Figure 3). The grid size was 3 mm, and the resistance was equal to 350 Ω for both strain gauge types. The acquisitions were performed at 100 Hz sampling frequency, a value sufficiently high to avoid missing signal information.

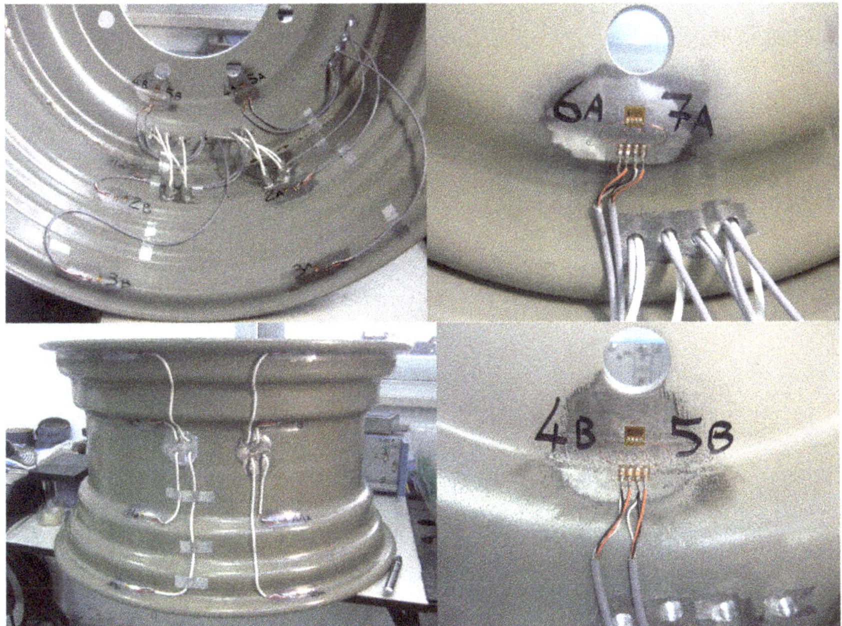

Figure 3. Strain gauges applied to the rim and folder of the wheel.

5. Analysis of the Time History Measured on the Wheel

Among all measured signals, the strain time histories measured on the outer rim of the wheel (strain gauge CH6 in Figure 4) were selected and analyzed in detail with the purpose of illustrating how the confidence interval of damage is constructed. In a preliminary stage, the analysis exploits some signal processing tools that are used to identify and describe some relevant features of the measured time history. Such tools can indicate, both qualitatively and quantitatively, whether the time history is nonstationary.

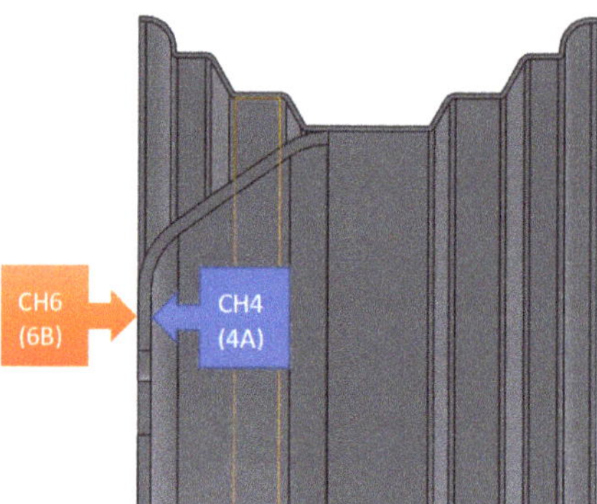

Figure 4. Location of strain gauge for channels CH4 (inner rim) and CH6 (outer rim). Reproduced from [12] with permission from Elsevier.

For the selected strain gauge channel, four time histories, corresponding to the four loading conditions mentioned in Section 4, are available. Each time history represents a well-defined state in which the wheel operates. On the other hand, for the confidence interval method in Section 3 to be applied, the individual time history $z(t)$ needs to be a single nonstationary switching signal formed by different states. For this reason, and for the sake of analysis, the four strain signals from channel CH6 were combined together to form a single nonstationary switching time history $z(t)$ with a total duration of $T = 3456$ s, see Figure 5a.

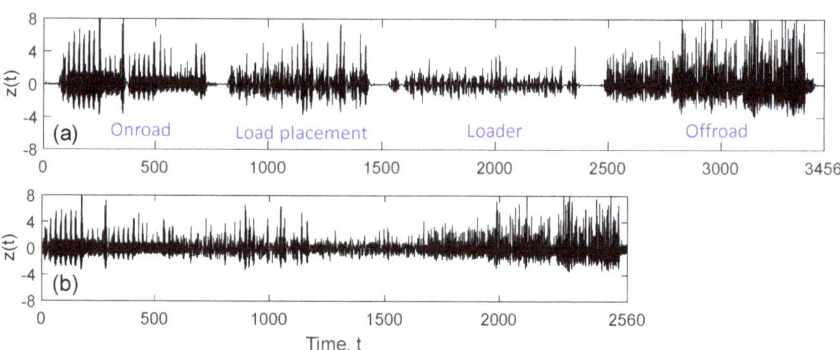

Figure 5. Measured time history used in the analysis (CH6, outer rim): (**a**) before and (**b**) after filtering and elimination of steady-state segments.

Nevertheless, this prior information about $z(t)$ was not exploited to decide a priori how many and which stationary states are present in $z(t)$. With the aim of best emphasizing the capabilities of the processing tools in detecting the relevant features of $z(t)$, the analysis proceeded with a sort of "blind approach" in which the only information available was the measured signal $z(t)$ itself.

In a first step, and without loss of generality, the single strain record obtained by measurements was normalized to a signal $z(t)$ with zero mean and unit variance (see Figure 5a). The time history $z(t)$ was very irregular over time t and had a marked nonstationary character.

Figure 5a highlights how the signal was formed by a few segments with zero or practically zero values, corresponding to the vehicle resting in a steady-state condition. These segments were not important for the subsequent analysis since they did not contribute with any fatigue cycle; therefore, they were eliminated by a trigger threshold algorithm applied directly to the original time history $z(t)$. In addition to this algorithm, a Butterworth filter with a cutoff frequency of 10 Hz was applied to eliminate high-frequency vibrations seen not to be of interest for the subsequent fatigue analysis. In fact, such a cutoff frequency value was higher than the range of frequencies that characterize the vehicle usage in testing and that are related to the low speed of vehicle and to lifting operations. After eliminating the steady-state segments and filtering the original signal, a shorter time history with duration $T = 2560$ s was obtained (see Figure 5b).

The time history $z(t)$ in Figure 5b was processed in the analysis stages described hereafter. It contained the relevant fatigue cycles formed by peaks and valleys, which were obtained by the measurement performed in the vehicle wheel during the duration $T = 3456$ s. As usual in structural durability, the entire time history was subjected to rainflow counting according to the "four-point algorithm" as per the ASTM E 1049 standard [28]; values were discretized into 64 bins. The three-dimensional plot of the rainflow matrix is displayed in Figure 6 as a histogram that shows the statistical distribution of amplitudes and mean values of rainflow cycles.

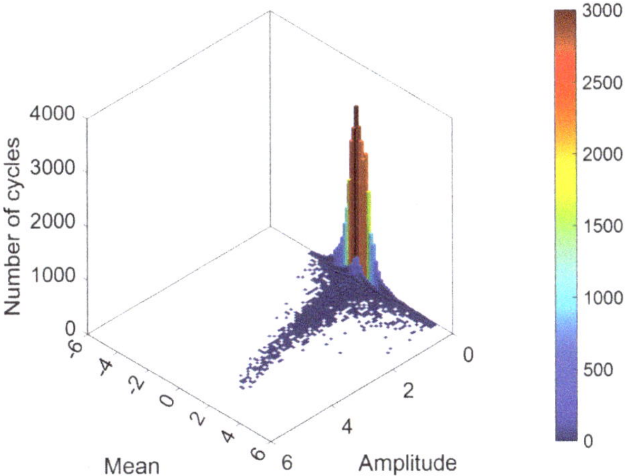

Figure 6. Three-dimensional histogram of rainflow matrix for the measured time history (CH6, outer rim), as a function of normalized amplitude and mean value of counted cycles.

Figure 6 evidences an asymmetric distribution around the zero mean value, with large-amplitude cycles having a positive mean value. Furthermore, Figure 6 demonstrates that the time history had a great deal of low-amplitude cycles with mean zero or close to zero. This trend suggests that the statistical distribution of mean values did not follow a Gaussian distribution, and that of amplitudes did not follow a Rayleigh distribution, as would be expected for a "regular" time history in which cycles have zero mean and symmetrical peak and valley; this type of random signal is often called "narrow-band" in the literature [29].

Although the three-dimensional rainflow matrix is an important characteristic of the signal, it collects in one single ensemble all the fatigue cycles counted in the same measured time history; therefore, it cannot be used to detect whether cycles counted in different portions of the time history follow, in fact, different statistical distributions. As a consequence, the nonstationarity of the measured time history cannot be verified by the rainflow matrix in Figure 6. To this end, other techniques (e.g., level-crossing spectra, cumulative spectra, short-time Fourier transform, run test) are applied to check the nonstationarity of the measured time history. They are grouped into two categories (qualitative and quantitative methods) depending on whether their results allow for conclusions based on statistical methods.

5.1. Qualifying the Nonstationarity

Since the confidence interval in Equation (7) is only applicable to a nonstationary time history formed by a sequence of stationary states, qualitative and quantitative methods were used to verify the nonstationarity of the measured time-history $z(t)$ in Figure 5b.

One qualitative method is based on the comparison of level-crossing spectra computed from different portions into which the whole signal has been divided. The level-crossing spectrum depicts the distribution of the number of times a signal upward crosses (upcrossings) a level [30], as a function of the level. The shape of the crossing spectrum is an indirect measure of the statistical properties of the random signal. The technique used here to check for the presence of a nonstationarity was to divide the time history into segments of equal length and to compare the level-crossing spectrum of each segment.

After dividing the measured time history $z(t)$ into three segments with the same length $T_s = 853.3$ s each, the level-crossing spectrum was computed for each segment; their comparison is illustrated in Figure 7.

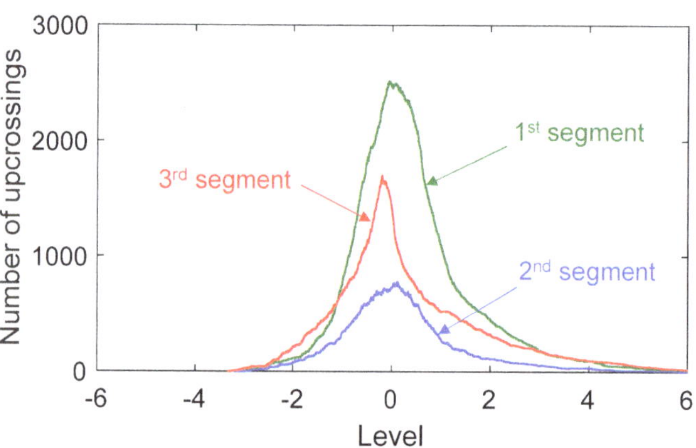

Figure 7. Comparison of level-crossing spectra from three segments of $z(t)$ with duration $T_s = 853.3$ s each (CH6 record, outer rim). Levels are normalized to signal standard deviation.

All three level-crossing spectra were not symmetric; rather, they had a positively skewed distribution characterized by a large number of crossings in the upper tail at positive values, with the presence of crossings even at extreme positive levels. This result is consistent with the presence of cycles with positive mean value observed in the rainflow matrix in Figure 6. In addition, it can be observed that all three level-crossing distributions in Figure 7 showed a similar shape but with a large difference in the number of upcrossings over the levels. This difference is a result in favor of nonstationarity, as it emphasizes the difference in load characteristics from one segment to another.

Another qualitative method to check for nonstationary features is the comparison of the amplitude distribution of rainflow cycles counted in different subsegments of $z(t)$ with same duration. As usual in structural durability, the amplitude distribution is given in terms of a cumulative (or loading) spectrum, which shows how the amplitude varies versus the number of cumulated cycles. The cumulative spectrum is also a useful tool to investigate the cumulative damage behavior [30].

Figure 8 compares three cumulative spectra obtained by first dividing the measured time history $z(t)$ into three segments of same duration $T_s = 853.3$, and then applying the rainflow counting method to each one. The comparison clearly highlights a marked difference in the amplitude distribution. It also reveals that a higher amplitude leads to worse agreement among cumulative spectra. This disagreement further suggests that the measured time history $z(t)$ is nonstationary.

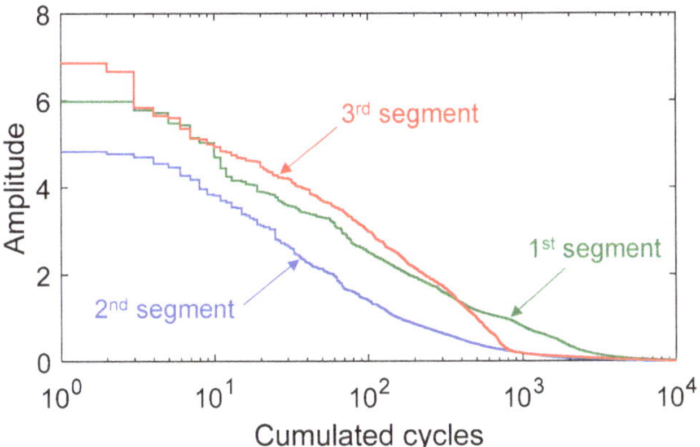

Figure 8. Comparison of cumulative spectra from three segments of $z(t)$ of same duration $T_s = 853.3$ s each (CH6 record, outer rim). Amplitude is normalized to signal standard deviation.

The short-time Fourier transform (STFT) is a method that can be used to analyze how the frequency content of a time history varies over time [31,32]. In order to apply the STFT, the time history must first be divided into a series of short time windows, and the Fourier transform is then computed for the signal within each window.

The method has some requirements to control the resolution achievable in the frequency domain and time domain. The frequency resolution can be improved by increasing the time length of the windows; conversely, the time resolution increases when the window length decreases. These opposite trends (linked to the Heisenberg–Gabor uncertainty principle [33]) emphasize that a high resolution cannot be achieved simultaneously in the time domain and frequency domain; a compromise must be found. On the other hand, taking a time history with fixed time duration T, the STFT plots (spectrograms) in both time and frequency domains are also altered using overlapped windows and zero padding [31]. A positive overlapping between consecutive segments, for example, yields a smoother spectrogram, although it does by no means decrease the estimation error, unlike Welch's method where overlapping improves the spectrum estimation accuracy [34]. It can indeed be demonstrated that, regardless of segment overlapping, the STFT has a normalized random error (coefficient of variation) of $\sqrt{4/\pi - 1} \cong 0.522$ [32], a value rather large as it means that the standard deviation of the estimate is about one-half of the value being estimated.

Here, the STFT was applied to the measured time history using two different window lengths ($T_w = 10$ s and $T_w = 40$ s) without overlapping and zero padding (see Figure 9). The various colors identify different spectrum levels. A look at the figures shows that, for both window lengths T_w, the frequency content of the signal had significant variations over time t. A marked change in the frequency content is apparent, which in the initial signal portion was mainly concentrated on a narrow range around 2 Hz, before shifting toward lower frequencies and eventually shifting again in the last signal portion to cover a broader frequency range up to 3 Hz. Note that this trend over time can be appreciated best with the spectrogram obtained by a shorter window, which indeed yields a better time resolution. The longer window, by contrast, allows for a better visualization of the frequency content.

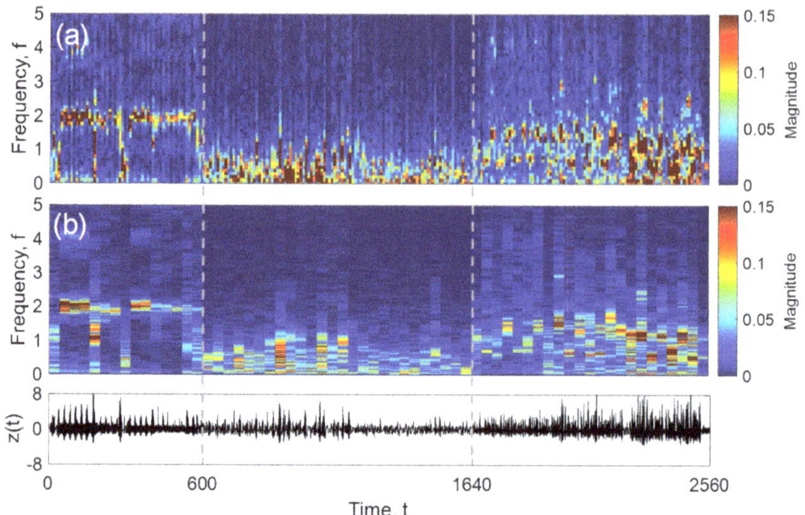

Figure 9. Short-time Fourier transform applied to the measured time history $z(t)$ with two window lengths (CH6 record, outer rim): (**a**) $T_w = 10$ s and (**b**) $T_w = 40$ s.

Regardless of window length, a characteristic time evolution clearly emerged. More precisely, Figure 9a,b suggest that the entire signal could approximately be divided into three distinct segments with different frequency content. The first segment can be identified visually from 0 to 600 s, the second from 600 s to 1640 s, and the last from 1640 s to 2560 s. This means that the entire measured time history $z(t)$ could be classified, at least visually, as a nonstationary signal formed by three quasi-stationary or nearly stationary states with different duration T_s. Surprisingly, this finding contrasts with the signal being in fact formed by four portions obtained in four different testing conditions. It may then be concluded that two of the portions had very similar frequency characteristics and could be grouped together for the subsequent derivation of the confidence interval. Albeit apparently contrasting, this result evidences the capabilities of the proposed processing technique for discriminating the various states in $z(t)$ solely on the basis of the knowledge of the signal itself, and without any prior knowledge on the testing conditions, which indeed may often not be available to the analyst.

By considering the outcome of all previous qualitative analyses, the measured time history from the vehicle wheel appears to be nonstationary. This conclusion, albeit quite plausible when also considering the nonstationary testing conditions in which the signal was obtained, was drawn only from a visual inspection of results and, therefore, needs to be supported by further conclusions based on quantitative methods, as considered in the next section.

5.2. Quantifying the Nonstationarity

This section considers a statistical method that can be used to draw conclusions on a quantitative basis on whether the measured time history is nonstationary. The statistical method is a nonparametric test called the Wald–Wolfowitz run test or simply run test [35]. The results obtained from this method are aimed at supporting the conclusions from the more qualitative analysis in the previous section.

The run test can detect whether a sequence of values is characterized by an underlying deterministic (not random) trend that leads to a nonstationary behavior. More specifically, the run test is a hypothesis test that checks for the null hypothesis "the sequence has no deterministic trend and is stationary". The run test is based on the definition of "run" as a sequence of identical observations followed and preceded by a different observation or no observation at all [35–37]. The sequence of values is usually classified into two dichotomic categories being above (+) or below (−) a reference value, usually the median of the sequence.

In order to apply the run test to a random time history that is a continuous function of time, a discrete sequence of values must be obtained. For this purpose, the time history is divided into segments and, for each segment, a statistical quantity (usually the root-mean-square (RMS) value) is computed. Accordingly, the time history is converted into a discrete sequence of RMS values, which is then tested for stationarity by the run test [38].

For a sequence with a large number of RMS values (i.e., more than 10), the distribution of runs in the sequence approaches a normal distribution with mean value $\mu_r = E[r]$ and variance $\sigma_r^2 = Var(r)$ [36,37]

$$\mu_r = 1 + n_+, \quad \sigma_r^2 = \frac{n_+(n_+ - 1)}{2n_+ - 1}, \tag{8}$$

where n_+ is the number of observations above the median. Equation (8) assumes that $n_+ = n_-$, i.e., the number of observations above and below the median is equal, a condition that holds true according to the definition of median for an even sequence of values (for an odd sequence, the value equal to the median is discarded, indeed).

In the run test, the acceptance region of the null hypothesis at the significance level β is

$$r_{n_+, 1-\beta/2} < r \le r_{n_+, \beta/2}, \tag{9}$$

in which $r_{n_+, 1-\beta/2}$ and $r_{n_+, \beta/2}$ are, respectively, the lower and upper limits of the region, which can be found in statistical tables [39] or determined (if the sequence has >10 values) by assuming a normal distribution of runs with μ_r and σ_r^2 in Equation (8).

On the basis of the outcome of the run test, a time history is classified as stationary if the number of runs r falls inside the acceptance region. Otherwise, the time history is classified as nonstationary.

The run test method was here applied to the measured time history $z(t)$ by taking two different values of the segment length ($T_s = 10$ s and $T_s = 20$ s; see Figure 10). The first value coincides with that already used for the STFT. Two different values were considered with the aim of investigating the sensitivity of the run test to the segment length. As already pointed out in [38], a long value of T_s has the effect of smoothing local variations in the RMS value, such that the time history tends to be classified as stationary. Conversely, a too small value of T_s will increase the appearance of the time history as nonstationary [38].

Figure 10a shows the results of the run test when the time segment length was $T_s = 10$ s. For $n_+ = n_- = 128$ observations above and below the median, and a 5% significance level, the upper and lower limits according to Equation (9) are $r_{128,0.975} = 113.3$ and $r_{128,0.025} = 144.6$. These limits were obtained by computing the mean and variance in Equation (8) and assuming a normal distribution of runs. The number of runs $r = 83$ was achieved by counting how many times the RMS values, computed in each segment, crossed up and down the median. Given that the observed number of runs $r = 83$ fell outside the acceptance region, the time history was classified as nonstationary by the run test.

The same conclusion was also obtained with a longer segment length (Figure 10b). For the case of $T_s = 20$ s, the number of observations above/below the median was $n_+ = n_- = 64$ and, for a 5% significance, the limits of the acceptance region are now $r_{64,0.975} = 54$ and $r_{64,0.025} = 76$. With a longer segment length, the number of runs decreased to $r = 35$, a value that was nevertheless outside the acceptance region for stationarity. Although this value is closer to the lower bound of the region than the previous case, the run test again yielded an indication in favor of nonstationarity.

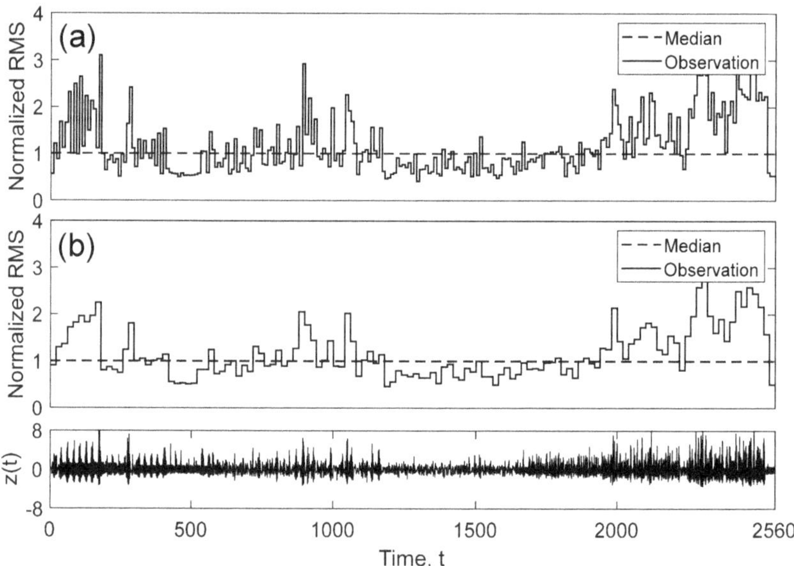

Figure 10. Run test method applied to the measured time history $z(t)$ with two segment lengths (CH6 record, outer rim): (**a**) $T_s = 10$ s and (**b**) $T_s = 20$ s.

According to [38], a "stationarity index" $\gamma_r = r/\mu_r$ is defined as the ratio of the observed number of runs to the expected number of runs for a stationary signal. In [38], the parameter γ_r was used as an indicator of the degree of stationarity, where a high value of γ_r indicates that a signal is "more stationary". For the above examples, the two values would be $\gamma_r = 83/129 = 64.3\%$ (for $T_s = 10$ s) and $\gamma_r = 35/65 = 53.8\%$ (for $T_s = 20$ s), indicating that a shorter segment length made the time history appear as more stationary.

6. Confidence Interval of Damage: Vehicle Wheel Data

After the previous analysis confirmed that the time history $z(t)$ measured in the industrial vehicle wheel (see Figure 5b) is nonstationary and also formed by a sequence of stationary states, it was possible to apply the confidence interval of the damage in Equation (7).

As a first step, the confidence interval required that the individual states in $z(t)$ be identified. On the basis of the results of the STFT analysis, the entire measured signal was divided into three segments ($N_S = 3$). Each segment was identified by the STFT spectrum in Figure 9. These segments may be classified as quasi-stationary or nearly stationary because they have a frequency content almost constant over time.

The confidence interval was constructed with a 95% confidence level by computing the sample values $D_{B,ij}(T_{B,i})$ and $\hat{\sigma}^2_{D_{B,i}}$, where $i = 1, 2, 3$ is the number of states and $j = 1, 2, \ldots N_B$ is the number of block subdivisions within each state. $D_{B,ij}(T_{B,i})$ is the fatigue damage in block j located within state i. The damage was calculated by assuming an S–N curve $S^m N = K$ with normalized strength constant $K = 1$ and inverse slope $m = 3$.

The confidence interval was then calculated by varying the number of blocks in the range $N_B = 2, 3 \ldots 10$. In the limit case $N_B = 10$, each block contained approximately 600 counted cycles, a number large enough to ensure the approximation of $D(T) \cong \sum_{i=1}^{N_S} \sum_{j=1}^{N_B} D_{B,ij}(T_{B,i})$ to hold, at least for the measured time history $z(t)$ considered in this study. This is confirmed by the results shown below.

Figure 11 displays the confidence intervals of the damage as a function of different number of block subdivisions, N_B. All damage values were normalized to the damage $D(T)$ of the whole time history $z(t)$. The intervals were constructed, according to Equation (7), around the total damage (solid circles) approximated by the double summation of the sample damage $D(T) \cong \sum_{i=1}^{N_S} \sum_{j=1}^{N_B} D_{B,ij}(T_{B,i})$.

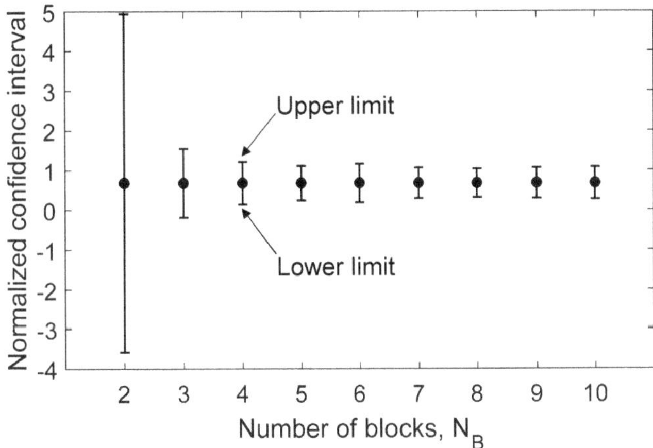

Figure 11. Confidence interval of the damage for the measured time history $z(t)$ from the vehicle wheel (CH6 record, outer rim), as a function of the number of blocks, N_B.

It is interesting to note that the damage value identified by the solid circle remained almost constant for any number of blocks, N_B. This result confirms that the block subdivision eliminated a very negligible number of cycles, such that the approximation $D(T) \cong \sum_{i=1}^{N_S} \sum_{j=1}^{N_B} D_{B,ij}(T_{B,i})$ was perfectly acceptable.

Figure 11 also confirms a general trend already observed in [21] for a stationary time history. The confidence interval became narrower as the number of blocks N_B increased. However, upon increasing N_B, the intervals did not approach zero; rather, they appeared to converge approximately to a sort of minimum interval width.

This outcome means that, while the prediction error for the damage was reduced insofar as the number of block damages (i.e., the block number N_B) increased, it could not become infinitely zero since the number of fatigue cycles was already given in the entire time history. On the other hand, the results shown in Figure 11 prove the importance of using as many blocks as possible in order to decrease the prediction error.

The trends of the confidence intervals here applied to the time history $z(t)$ measured in the industrial vehicle wheel agree with the findings obtained in previous studies with simulations and experiments using mountain bike data [21,22].

Nevertheless, while it was possible for the latter results to be verified by comparison with the expected damage $E[D(T)]$ (which may or may not fall within the confidence interval), this verification was not achievable in the practical case study analyzed here, because the expected damage $E[D(T)]$ was not available.

This circumstance is the rule in practice. Indeed, the expected damage would correspond to the average damage computed over an infinite ensemble of time histories acting on the vehicle wheel, which for obvious reasons was not available. Nevertheless, from the positive feedback from previous validation studies on the same approach presented here, there is reason to believe that the proposed confidence interval approach is in fact correct. On the other hand, in practical applications, it is common to have available only one single measured time history, as in the case study discussed so far in this work.

As a final remark, if the unknown expected damage $E[D(T)]$ were underestimated by the observed value of $D(T)$, a structure or component would be designed unsafely. Only for damage values greater than $E[D(T)]$ would the structure or component be in the safe region. On the other hand, in all such practical cases (as that analyzed in the present work) in which $E[D(T)]$ is unknown, it is not possible to establish a priori how far $D(T)$ is from $E[D(T)]$. In this regard, the confidence interval, which includes information about the variability of $D(T)$, may help in drawing meaningful conclusions on this point. More specifically, it is suggested to take the upper limit as the reference damage value to be used in design [21].

7. Analysis of Time History from Inner Wheel Rim

In addition to presenting results, the purpose of Section 6 was also to illustrate in great detail the various analysis steps of an individual nonstationary time history. As an example, the procedure was applied to the time history from the outer wheel rim (CH6 in Figure 4). For comparison, the procedure was also applied to the inner rim time history (CH4). Most of the analysis details already discussed in Section 6 are now skipped, thus focusing the attention more on the results.

At first glance, the CH4 time history appears to be very similar to the CH6 record depicted in Figure 5. Upon a closer look, however, it is possible to observe that rainflow cycles were distributed more symmetrically around the global mean value (see Figure 12a). Compared to the distribution in Figure 6, the large-amplitude cycles had a mean value zero or closer to zero. The level-crossing spectra computed in three consecutive segments of the CH4 record are displayed in Figure 12b; they are very similar to those of CH6 shown in Figure 7, apart from the lower number of crossings counted. Interesting in Figure 12b is the peculiar two-spike shape of the level-crossing spectrum in the first segment of the CH4 time history.

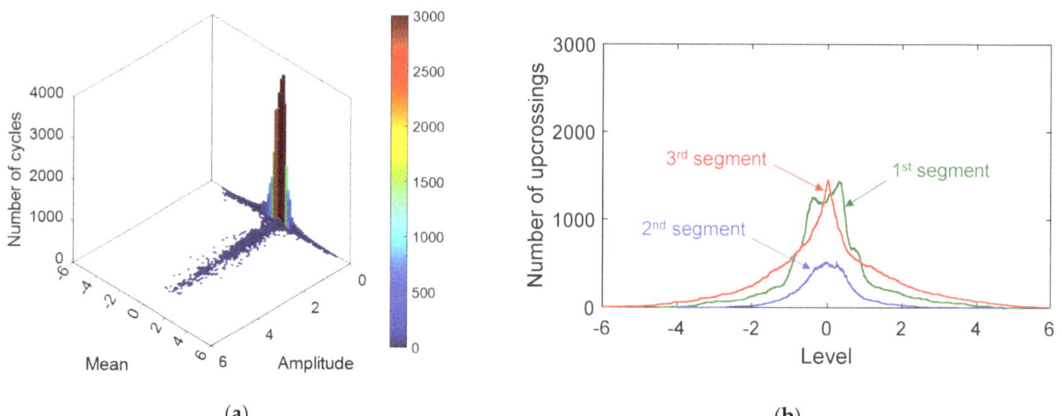

Figure 12. CH4 record, inner rim: (**a**) Rainflow matrix; (**b**) level-crossing spectrum. Amplitude, mean and level are normalized to signal standard deviation.

As with the CH6 record, in a preliminary stage, the CH4 time history was also analyzed by the short-time Fourier transform and the run test to detect nonstationary features. As a representative example, Figure 13 shows the results obtained using both techniques when choosing a time window of 10 s. The output for the CH4 time history is practically coincident with that already pointed out for the CH6 signal. Not only did both techniques suggest, visually and quantitatively, that the signal was in fact nonstationary, but they also allowed the signal to be separated for the subsequent analysis into three segments with similar characteristics.

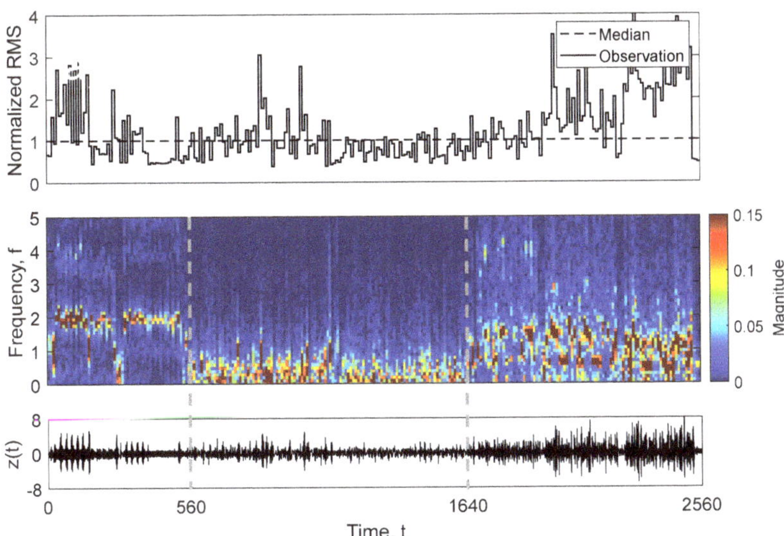

Figure 13. Short-time Fourier transform and run test applied to CH4 time history (inner rim), shown at the bottom of the figure. In both methods, the window time length was 10 s.

Once the individual CH4 time history was divided into three states, and then into blocks, the confidence interval of damage was computed as a function of the number of blocks (see Figure 14). Without a doubt, the results and trends for the CH4 signal are pretty much similar to those obtained for the CH6 time history (see Figure 11).

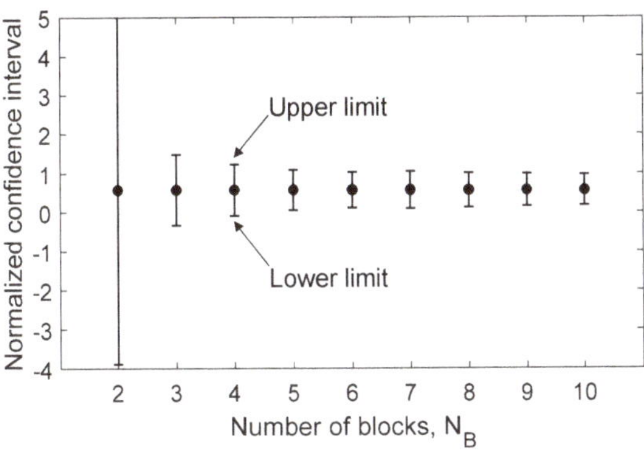

Figure 14. Confidence interval of the damage for the measured time history $z(t)$ from the vehicle wheel (CH4 record, inner rim), as a function of number of blocks, N_B

8. Conclusions

This article presented an approach for analyzing a single nonstationary random fatigue loading with the purpose of constructing the confidence interval of the expected fatigue damage. Specifically, the approach was applied to a so-called switching nonstationary loading formed by a finite number of stationary or pseudo-stationary states. The obtained confidence interval is a statistical tool to assess the sampling variability of a single fatigue damage value computed from only one individual time history.

The approach was applied here to the strain time histories measured in a wheel of a telescopic handler vehicle when subjected to four testing conditions. The time histories were recorded in the inner and outer wheel rim. In a preliminary phase, qualitative and quantitative methods (short-time Fourier transform, level-crossing spectrum, run test) were used to deduce some characteristic features of the signal. For example, they allowed deciding how many and which states the whole signal was divided into before calculating the confidence interval. After the loading was divided into states, and each state was further subdivided into smaller portions (blocks), the confidence interval could eventually be determined.

The obtained results allowed reaching the following conclusions:

- The outcomes of both qualitative and quantitative methods suggested that the individual measured signals could in fact be classified as nonstationary. This conclusion was confirmed independently of the choice of the length of the time window used by the short-time Fourier transform (STFT) and by the run test;
- On the basis of the outcome from the STFT, the original single nonstationary time history was divided into three different states. Surprisingly, this finding contrasts with the fact that the signal was, in fact, composed by four consecutive signal portions corresponding to four different loading conditions. It may then be concluded that two of the portions had very similar frequency characteristics and could be grouped together for the subsequent derivation of the confidence interval;
- Subdivision into states and then blocks determined a loss of a very small fraction of fatigue cycles, i.e., those formed by peaks and valleys falling into different states or blocks before subdivision. This small loss by no means affected the accuracy of the calculated confidence interval;
- As the number of blocks increased, the width of the confidence interval decreased, albeit not indefinitely, since it eventually attained approximately a sort of minimum value. This trend confirmed that the prediction error on the expected damage was reduced by increasing the number of blocks, suggesting the use of as many blocks as possible. On the other hand, the fact that the confidence interval width did not approach zero also highlighted that an increase only in the number of blocks would not yield any additional information on the cycle distribution to allow for a further increase in prediction accuracy, i.e., the amount of information was established by the set of fatigue cycles characterizing the given individual time history;
- In conclusion, it is suggested to use the upper limit of the confidence interval of damage as a design value in a structural durability assessment.

Author Contributions: Conceptualization, J.M.E.M. and D.B.; experiments, L.S.; data analysis, J.M.E.M.; writing—original draft preparation, J.M.E.M.; writing—review and editing, D.B. and L.S.; supervision, D.B. All authors have read and agreed to the published version of the manuscript.

Funding: The research activity of one author (J.M.E.M.) was partially funded by the CTU Global Postdoc Fellowship Program (research topic #2-11).

Data Availability Statement: Data are contained within the article.

Acknowledgments: The authors would like to thank Eng. Gianpietro Bramè of Moveero Ltd. company for the information provided during the development of this research.

Conflicts of Interest: The authors declare no conflict of interest.

References

1. Balkwill, J. Weight transfer and wheel loads. In *Performance Vehicle Dynamics: Engineering and Applications*; Elsevier: Amsterdam, The Netherlands, 2018. [CrossRef]
2. Peeters, M.; Kloster, V.; Fedde, T.; Frerichs, L. Integrated wheel load measurement for tractors. *Landtechnik* **2018**, *73*, 116–128. [CrossRef]
3. Muhlmeier, M. Evaluation of wheel load fluctuations. *Int. J. Veh. Des.* **1995**, *16*, 397–411. [CrossRef]
4. Solazzi, L. Wheel rims for industrial vehicles: Comparative experimental analyses. *Int. J. Heavy Veh. Syst.* **2011**, *18*, 214–225. [CrossRef]
5. Xiaoyin, W.; Xiandong, L.; Yingchun, S.; Xiaofei, W.; Wanghao, L.; Yue, P. Lightweight design of automotive wheel made of long glass fiber reinforced thermoplastic. *Proc. Inst. Mech. Eng. Part C J. Mech. Eng. Sci.* **2016**, *230*, 1634–1643. [CrossRef]

6. Wilczynski, A.; Bartczak, M.; Siczek, K.; Kubiak, P. Carbon fibre reinforced wheel for fuel ultra–efficient vehicle. *Mech. Mech. Eng.* **2018**, *22*, 1419–1437. [CrossRef]

7. Solazzi, L. Applied research for weight reduction of an industrial trailer. *FME Trans.* **2012**, *40*, 57–62.

8. Collotta, M.; Solazzi, L. New design concept of a tank made of plastic material for firefighting vehicle. *Int. J. Automot. Mech. Eng.* **2017**, *14*, 4603–4615. [CrossRef]

9. Solazzi, L.; Assi, A.; Ceresoli, F. Excavator arms: Numerical, experimental and new concept design. *Compos. Struct.* **2019**, *217*, 60–74. [CrossRef]

10. Solazzi, L.; Scalmana, R. New design concept for a lifting platform made of composite material. *Appl. Compos. Mater.* **2013**, *20*, 615–626. [CrossRef]

11. Solazzi, L. Experimental and analytical study on elevating working platform. *Procedia Eng.* **2017**, *199*, 2597–2602. [CrossRef]

12. Cima, M.; Solazzi, L. Experimental and analytical study of random fatigue, in time and frequencies domain, on an industrial wheel. *Eng. Fail. Anal.* **2021**, *120*, 105029. [CrossRef]

13. Mazzoni, A.; Solazzi, L. Experimental field test on a multipiece steel wheel and influence of the material properties on its fatigue life evaluation. *Eng. Fail. Anal.* **2022**, *135*, 106106. [CrossRef]

14. Bendat, J.S. Probability Functions for Random Responses: Prediction of Peaks, Fatigue Damage, and Catastrophic Failures. NASA-CR-33, 1 April 1964. Available online: https://ntrs.nasa.gov/api/citations/19640008076/downloads/19640008076.pdf (accessed on 31 January 2022).

15. Mark, W.D. The Inherent Variation in Fatigue Damage Resulting from Random Vibration. Ph.D. Thesis, Department of Mechanical Engineering, Massachusetts Institute of Technology (M.I.T), Cambridge, MA, USA, 1961.

16. Crandall, S.H.; Mark, W.D.; Khabbaz, G.R. The variance in Palmgren-Miner damage due to random vibration. In Proceedings of the 4th US National Congress of Applied Mechanics, Berkeley, CA, USA, 18–21 June 1962; Rosenberg, R.M., Ed.; American Society of Mechanical Engineers (ASME): New York, NY, USA, 1962; Volume 1, pp. 119–126.

17. Madsen, H.O.; Krenk, S.; Lind, N.C. *Methods of Structural Safety*; Prentice-Hall: Hoboken, NJ, USA, 1986.

18. Low, Y.M. Variance of the fatigue damage due to a Gaussian narrowband process. *Struct. Saf.* **2012**, *34*, 381–389. [CrossRef]

19. Low, Y.M. Uncertainty of the fatigue damage arising from a stochastic process with multiple frequency modes. *Probab. Eng. Mech.* **2014**, *36*, 8–18. [CrossRef]

20. Marques, J.M.E.; Benasciutti, D. Variance of the fatigue damage in non-Gaussian stochastic processes with narrow-band power spectrum. *Struct. Saf.* **2021**, *93*, 102131. [CrossRef]

21. Marques, J.M.E.; Benasciutti, D.; Tovo, R. Variability of the fatigue damage due to the randomness of a stationary vibration load. *Int. J. Fatigue* **2020**, *141*, 105891. [CrossRef]

22. Marques, J.M.E. Confidence intervals for the expected damage in random loadings: Application to measured time-history records from a Mountain-bike. *IOP Conf. Ser. Mater. Sci. Eng.* **2021**, *1038*, 012025. [CrossRef]

23. Marques, J.M.E.; Benasciutti, D. Evaluating confidence interval of fatigue damage from one single measured non-stationary time-history. In Proceedings of the VIRTUAL 20th International Colloquium on Mechanical Fatigue of Metals, Wrocław, Poland, 15–17 September 2021.

24. Benasciutti, D.; Marques, J.M.E. Methods for estimating the statistical variability of the fatigue damage computed in one single stationary or non-stationary random time-history record. In Proceedings of the 3rd International Conference on Structural Integrity for Offshore Energy Industry (STRUCTURAL INTEGRITY—2021), Aberdeen, UK, 15–16 November 2021; ASRANet Ltd.: Glasgow, UK, 2021; pp. 28–37, ISBN 978-1-8383226-3-2.

25. *ASTM E739–10 (2015)*; Standard Practice for Statistical Analysis of Linear or Linearized Stress-Life (S-N) and Strain-Life (ε-N) Fatigue Data. ASTM International: West Conshohocken, PA, USA, 2017. Available online: https://www.astm.org/e0739-10r15.html (accessed on 10 February 2022).

26. Johannesson, P. Rainflow cycles for switching processes with Markov structure. *Probab. Eng. Inform. Sci.* **1998**, *12*, 143–175. [CrossRef]

27. Montgomery, D.C.; Runger, G.C. *Applied Statistics and Probability for Engineers*, 6th ed.; John Wiley & Sons: Hoboken, NJ, USA, 2014.

28. *ASTM E1049–85*; Standard Practices for Cycle Counting in Fatigue Analysis. ASTM International: West Conshohocken, PA, USA, 2017. Available online: https://www.astm.org/c1049-85r17.html (accessed on 10 February 2022).

29. Benasciutti, D. *Fatigue Analysis of Random Loadings: A Frequency Domain Approach*; LAP Lambert Academic Publishing: Saarbrücken, Germany, 2012; ISBN 9783659123702.

30. Johannesson, P.; Speckert, M. *Guide to Load Analysis for Durability in Vehicle Engineering*; John Wiley & Sons: Hoboken, NJ, USA, 2014.

31. Cohen, L. *Time-Frequency Analysis*; Prentice-Hall: Hoboken, NJ, USA, 1995.

32. Bendat, J.S.; Piersol, A.G. *Random Data: Analysis and Measurement Procedures*, 4th ed.; John Wiley & Sons: Hoboken, NJ, USA, 2010.

33. Mallat, S. *A Wavelet Tour of Signal Processing: The Sparse Way*, 3rd ed.; Academic Press: Cambridge, MA, USA, 2009.

34. Welch, P.D. The use of Fast Fourier Transform for the estimation of power spectra: A method based on time averaging over short, modified periodograms. *IEEE Trans. Audio Electroacoust.* **1967**, *15*, 70–73. [CrossRef]

35. Wald, A.; Wolfowitz, J. On a test whether two samples are from the same population. *Ann. Math. Stat.* **1940**, *11*, 147–162. [CrossRef]

36. Hald, A. *Statistical Theory with Engineering Applications*; John Wiley & Sons: Hoboken, NJ, USA, 1955.
37. Mood, A.M. The distribution theory of runs. *Ann. Math. Stat.* **1940**, *11*, 367–392. [CrossRef]
38. Rouillard, V. Quantifying the non-stationarity of vehicle vibrations with the run test. *Packag. Technol. Sci.* **2014**, *27*, 203–219. [CrossRef]
39. Swed, F.S.; Eisenhart, C. Tables for testing randomness of grouping in a sequence of alternatives. *Ann. Math. Stat.* **1943**, *14*, 66–87. [CrossRef]

Review

An Overview of Fatigue Testing Systems for Metals under Uniaxial and Multiaxial Random Loadings

Julian M. E. Marques [1,*], Denis Benasciutti [1,*], Adam Niesłony [2] and Janko Slavič [3]

[1] Department of Engineering, University of Ferrara, Via Saragat 1, 44122 Ferrara, Italy
[2] Department of Mechanics and Machine Design, Faculty of Mechanical Engineering, Opole University of Technology, Prószkowska 76, 45-758 Opole, Poland; a.nieslony@po.edu.pl
[3] Faculty of Mechanical Engineering, University of Ljubljana, Aškerčeva 6, 1000 Ljubljana, Slovenia; janko.slavic@fs.uni-lj.si
* Correspondence: nzvjnm@unife.it (J.M.E.M.); denis.benasciutti@unife.it (D.B.); Tel.: +39-0532-974104 (J.M.E.M.); +39-0532-974976 (D.B.)

Abstract: This paper presents an overview of fatigue testing systems in high-cycle regime for metals subjected to uniaxial and multiaxial random loadings. The different testing systems are critically discussed, highlighting advantages and possible limitations. By identifying relevant features, the testing systems are classified in terms of type of machine (servo-hydraulic or shaker tables), specimen geometry and applied constraints, number of load or acceleration inputs needed to perform the test, type of loading acting on the specimen and resulting state of stress. Specimens with plate, cylindrical and more elaborated geometry are also considered as a further classification criterion. This review also discusses the relationship between the applied input and the resulting local state of stress in the specimen. Since a general criterion to classify fatigue testing systems for random loadings seems not to exist, the present review—by emphasizing analogies and differences among various layouts—may provide the reader with a guideline to classify future equipment.

Keywords: fatigue; testing systems; random loadings; servo-hydraulic; shaker table

Citation: Marques, J.M.E.; Benasciutti, D.; Niesłony, A.; Slavič, J. An Overview of Fatigue Testing Systems for Metals under Uniaxial and Multiaxial Random Loadings. *Metals* **2021**, *11*, 447. https://doi.org/10.3390/met11030447

Academic Editor: Tilmann Beck

Received: 30 January 2021
Accepted: 3 March 2021
Published: 8 March 2021

Publisher's Note: MDPI stays neutral with regard to jurisdictional claims in published maps and institutional affiliations.

1. Introduction

Mechanical components are often subjected to random loadings during their service life. Due to these loads, components may be exposed to a local random stress, which can be uniaxial (i.e., only one stress component) or multiaxial (i.e., two or more stress components). To estimate the component life, engineers usually perform a structural durability assessment in the predesign stage, often with the aid of a finite element (FE) analysis.

If the nodal random stress is uniaxial, the approach commonly followed makes use of rainflow counting and the Palmgren–Miner rule to compute the damage of the nodal stress output, based on uniaxial strength data given as an S–N curve. This analysis can be developed in time domain or, equivalently, in frequency domain [1].

If, instead, the nodal stress in the FE model is multiaxial, the analysis requires the use of a multiaxial fatigue criterion, which can also be formulated in time domain or frequency domain [2]. By analyzing all nodal results of a FE model (e.g., hundreds of thousands), the computation time is very long using a time-domain criterion. In addition, it may become impracticable when a huge number of planes have to be scanned in the whole three-dimensional FE model in the case of multiaxial fatigue criteria using the critical plane concept [3–5]. In general, for both the uniaxial and multiaxial cases, frequency domain solutions are several orders of magnitude faster than time domain simulations [6–9].

Methods for durability analysis, either in time domain or frequency domain, have to rely on material strength data obtained by experimental laboratory tests. Such tests are performed by applying loadings on a specimen, where cracks nucleate and grow until the specimen breaks. Tests with simple loadings (e.g., axial, bending and torsion) are carried out to estimate the material fatigue strength data in the uniaxial case. These uniaxial tests yield an S–N curve in high-cycle fatigue regime, which characterizes the material strength behavior in terms of amplitudes versus number of cycles to failure. Uniaxial test data are also necessary for calibrating a multiaxial criterion. Once calibration has been carried out, specific tests with multiaxial loadings (e.g., bending and torsion) are also performed to gather the necessary data for validating the multiaxial criterion. It is, nevertheless, clear that systems and methodologies for fatigue testing play a paramount role in the durability analysis of both uniaxial and multiaxial states of stress.

Fatigue testing methodologies may vary, for instance, in terms of type of machine used. Two different types (servo-hydraulic or electrodynamic shaker tables) are usually adopted in laboratories. Although, in general, servo-hydraulic machines can be used with both plate and cylindrical specimens, they are normally employed with cylindrical specimens in the random fatigue testing methodologies addressed in this paper. By contrast, shaker tables also adopt specimen geometries other than cylindrical, for example, plate or more elaborated ones.

Various specimen geometries and layouts are considered in fatigue tests, too. While the use of servo-hydraulic machines or shaker tables sometimes leads to rather simple testing layouts or state of stress (e.g., uniaxial), in some cases, the testing systems are all but obvious. As an example, a widely used system layout is that considering a cantilever plate specimen with rectangular shape [10–14]. Mounted on shaker tables and excited at its base, this system produces a bending random loading, which results in a near uniaxial state of stress in the critical location (e.g., notch or hole). However, using a different system with a cantilever cylindrical specimen, it is possible to reach a biaxial state of stress with both normal and shear stresses [15]. Applying loads at the free end of the specimen by a uniaxial shaker allows the system only to develop a coupled bending–torsion random loading. To overcome this limitation, bending and torsion can be applied by two independent uniaxial shakers [15]. In this case, not only coupled (correlated) but also uncoupled (uncorrelated) bending–torsion random loadings are achieved.

Another interesting system with a cantilever cylindrical specimen is that using two masses of different weights fixed at the free end [16]. Excited by a uniaxial shaker at its base, this specimen experiences a bending–torsion loading that produces coupled normal and shear stress components. Inspired by this layout, a new testing system was designed to apply only bending or torsion, as well as coupled or uncoupled bending–torsion loadings by a tri-axis shaker [17–19]. This system permits the intensity and phase shift of bending and torsion loadings to be controlled independently, which results in local normal and shear stresses with any degree of correlation. Experimental and numerical analyses confirmed the system's behavior.

A more elaborated specimen with Y-shaped geometry was also proposed not only to accelerate fatigue tests by means of shaker tables but also to simulate a real-world scenario of complex structures [20–25]. Although this special Y-shaped system is excited by two uncoupled random loadings (in vertical and horizontal directions), it develops a biaxial state of stress at the critical location in terms of normal stresses.

As may be expected, and perhaps become more apparent in the following sections, performing tests with a uniaxial stress is much easier than executing tests with a biaxial stress, even if one makes use of a bending–torsion servo-hydraulic machine. The degree of complexity of the testing system increases the most when the biaxial state of stress is random. While a uniaxial stress can be achieved simply by a servo-hydraulic machine with axially loaded specimens, or by a uniaxial shaker table with a cantilever specimen, fatigue tests in a biaxial random state of stress require one or more machines (e.g., one tri-axis shaker or two uniaxial shakers), one or more types of input as force/torque or acceleration, a specific testing layout and/or a particular shape of specimen. Although it is true that carrying out a fatigue test with a biaxial state of stress is not as simple as executing a test with uniaxial stress, the interest in biaxial random fatigue tests has increased in the last decade in the scientific community [15,17–25].

By considering this increased interest, the present paper aims to review the random loading fatigue testing systems available in the literature. The different testing systems are critically discussed, highlighting advantages and possible disadvantages. Some general features are also identified, which allow testing systems to be classified and grouped in terms of type of testing machine, specimen geometry, applied constraints, type and among of input needed to carry out a fatigue test, type of random loadings acting on the specimen and the resulting local state of stress. Regarding the loadings applied on specimens in fatigue tests, this paper focuses on axial, bending, torsion, axial-torsion and bending–torsion loadings. Specimens subjected to such loadings yield a uniaxial or a biaxial state of stress, which is also discussed hereafter.

Finally, it must be emphasized that, although testing systems for fatigue random loading are reviewed in detail throughout the text, the relationship between experimental results and estimations by various multiaxial fatigue criteria—though interesting—is not the scope of the present paper. For a discussion on this topic, the reader may refer to [26,27].

2. Common Random Loadings in Fatigue Tests and Resulting State of Stress

A multiaxial state of stress at a critical point of a mechanical component is represented by a tensor $\sigma(t)$, which in the most general case has six independent stress components [28]. For this reason, the six independent stress components are conveniently arranged into a six-dimensional vector $s(t) = [\sigma_{xx}(t),\ \sigma_{yy}(t),\ \sigma_{zz}(t),\ \tau_{xy}(t),\ \tau_{yz}(t),\ \tau_{zx}(t)]$, where $\sigma_{xx}(t)$, $\sigma_{yy}(t)$ and $\sigma_{zz}(t)$ are normal stresses and $\tau_{xy}(t)$, $\tau_{yz}(t)$ and $\tau_{zx}(t)$ are the shear stresses in an X–Y–Z cartesian coordinate system.

Usually, fatigue cracks nucleate at the surface of mechanical components, where the local state of stress is biaxial or even uniaxial. Therefore, in laboratory fatigue tests, the aim is to replicate in a specimen the same biaxial or uniaxial state of stress, in which only two or less normal stress components are nonzero [26,28]. While, in plane stress, a biaxial stress may have up to three nonzero components, $\sigma_{xx}(t)$, $\sigma_{yy}(t)$ and $\tau_{xy}(t)$, special cases often employed in laboratory tests consider one normal $\sigma_{xx}(t)$ and one shear $\tau_{xy}(t)$ stress, or nonzero normal stresses in two directions $\sigma_{xx}(t)$, $\sigma_{yy}(t)$. The uniaxial cases frequently adopted refer to a pure normal stress $\sigma_{xx}(t)$ or shear stress $\tau_{xy}(t)$; see Figure 1.

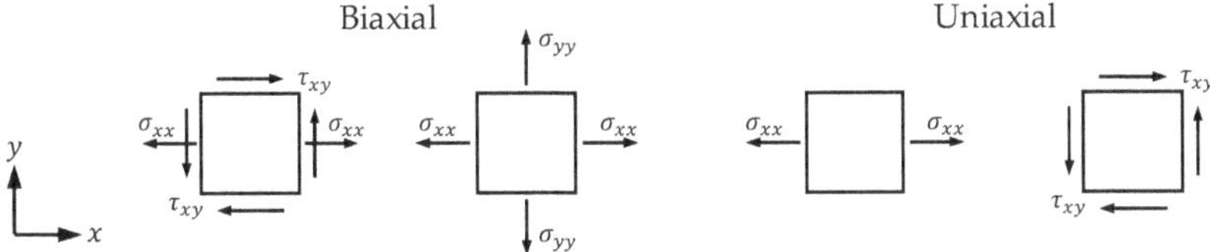

Figure 1. Common state of stress in fatigue testing: biaxial (e.g., normal and shear stresses or normal stresses in two directions); uniaxial (e.g., pure normal stress or pure shear stress).

In tests with constant amplitude loadings, it is common to use harmonic (sinusoidal) functions. For example, a biaxial normal shear stress with zero-mean is:

$$\sigma_{xx}(t) = \sigma_{xx,a}\sin(\omega t),\quad \tau_{xy}(t) = \tau_{xy,a}\sin(\omega t - \varphi) \tag{1}$$

where ω is the angular frequency and $\sigma_{xx,a}$ and $\tau_{xy,a}$ are the stress amplitudes.

For this biaxial state of stress (as, in fact, in any multiaxial case), the magnitude of stress components may change proportionally (in-phase) or nonproportionally (out-of-phase) with time. In particular, the ratio between normal and shear stresses, $\sigma_{xx}(t)/\tau_{xy}(t)$, at any time instant does not vary if $\varphi = 0$, and both stresses follow two in-phase harmonic functions. Instead, the ratio varies with time if $\varphi \neq 0$ and the two harmonic functions are out-of-phase. Additionally, the orientation of principal stress directions may change or not, depending on the value of φ: for in-phase stresses, they remain fixed; for out-of-phase stress, they change with time.

For harmonic stresses with the same frequency, the phase shift φ provides a simple measure of the degree of nonproportionality. This definition cannot be extended to random stresses, as they can be viewed as a superposition of many harmonic functions with different frequencies and phase shifts. In the random case, a statistical approach is needed to express the concepts of "fully correlated", "partially correlated" or "not correlated" (uncorrelated) signals. To this end, one introduces the so-called correlation coefficient between two random stresses, which is defined as the covariance normalized to the standard deviation (see below).

In tests with multiaxial random loadings, each stress component is a zero-mean stationary Gaussian process, which can be characterized in the frequency domain by a Power Spectral Density (PSD) matrix. For a biaxial random stress, $\sigma_{xx}(t)$, $\sigma_{yy}(t)$ and $\tau_{xy}(t)$, the PSD matrix takes the form [9,29]:

$$S(\omega) = \begin{bmatrix} S_{xx,xx}(\omega) & S_{xx,yy}(\omega) & S_{xx,xy}(\omega) \\ S^*_{xx,yy}(\omega) & S_{yy,yy}(\omega) & S_{yy,xy}(\omega) \\ S^*_{xx,xy}(\omega) & S^*_{yy,xy}(\omega) & S_{xy,xy}(\omega) \end{bmatrix}, \quad S_{ij}(\omega) = \int_{-\infty}^{\infty} R_{ij}(\delta)e^{-i\omega\delta}\,\mathrm{d}\delta \quad (2)$$

in which $R_{ij}(\delta)$ is the autocorrelation (for $i = j$) and cross-correlation (for $i \neq j$) function in time-domain, and δ is a time lag. The diagonal terms of $S(\omega)$ are the auto-PSDs $S_{xx,xx}(\omega)$, $S_{yy,yy}(\omega)$ and $S_{xy,xy}(\omega)$, whereas the out-of-diagonal terms are the cross-PSDs $S^*_{xx,yy}(\omega)$, $S^*_{xx,xy}(\omega)$ and $S^*_{yy,xy}(\omega)$, where the superscript * denotes the complex conjugate. Hence, the cross-PSDs are the summation of a real and an imaginary part. The real part is an even function of ω (coincident spectrum or co-spectrum), and the imaginary part is an odd function of ω (quadrature spectrum or quad-spectrum). The PSD matrix in Equation (2) is Hermitian.

Thanks to the relationship between covariance terms and the zero-order spectral moments, $C_{ij} = \lambda_{0,ij}$, the covariance matrix can be defined as [29]:

$$C = \begin{bmatrix} C_{xx,xx} & C_{xx,yy} & C_{xx,xy} \\ C_{yy,xx} & C_{yy,yy} & C_{yy,xy} \\ C_{xy,xx} & C_{xy,yy} & C_{xy,xy} \end{bmatrix}, \quad C_{ij} = \int_{-\infty}^{\infty} S_{ij}(\omega)\,\mathrm{d}\omega \quad (3)$$

Equation (3) is a symmetric matrix. The main diagonal terms are the variance of each stress component, $C_{xx,xx} = \mathrm{Var}(\sigma_{xx}(t))$, $C_{yy,yy} = \mathrm{Var}(\sigma_{yy}(t))$ and $C_{xy,xy} = \mathrm{Var}(\tau_{xy}(t))$; the out-of-diagonal terms are the covariance of two components, $C_{xx,yy} = \mathrm{Cov}(\sigma_{xx}(t),\sigma_{yy}(t))$, $C_{xx,xy} = \mathrm{Cov}(\sigma_{xx}(t),\tau_{xy}(t))$ and $C_{yy,xy} = \mathrm{Cov}(\sigma_{yy}(t),\tau_{xy}(t))$. The covariances are used to define the correlation coefficient $r_{ij} = C_{ij}/\sqrt{C_{ii}C_{jj}}$ between stress components i and j. This coefficient is close to unity when two components are "fully correlated" (proportional); it tends to zero when they are "uncorrelated" (nonproportional) [29].

In some testing layouts, the type of local random stress in the specimen (e.g., biaxial normal and shear stresses $\sigma_{xx}(t)$, $\tau_{xy}(t)$) closely depends on the type of machine, specimen geometry, constrains, and type of excitation. For example, a cantilever cylindrical specimen with a circumferential notch can be subjected to various independent types of loadings applied at the free end, see Figure 2. In the most general case, they are axial load $P(t)$; torsion $M_T(t)$; and two bending moments, $M_{B,x}(t)$ and $M_{B,y}(t)$,, in two orthogonal planes. For this specimen geometry, the maximum stresses are located on the surface at the center of the notch.

Assuming a bending–torsion loading, $M_{B,x}(t)$ and $M_T(t)$, without $P(t)$ and $M_{B,y}(t)$, the cylindrical specimen experiences a biaxial normal and shear stress at the critical location. By contrast, the specimen under the two bending moments $M_{B,x}(t)$ and $M_{B,y}(t)$ develops a uniaxial normal stress at the notch surface. The location of the maximum normal stress changes or not, depending on whether the instantaneous value $M_{B,y}(t)/M_{B,x}(t)$ varies or not.

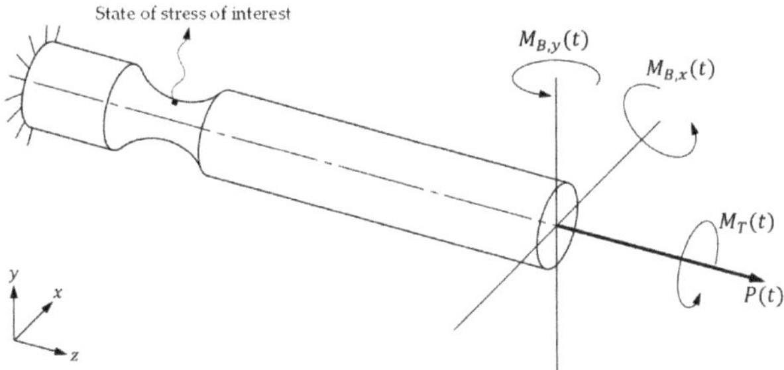

Figure 2. Notched cantilever cylindrical specimen under axial $P(t)$; torsion $M_T(t)$; and two bending moments, $M_{B,x}(t)$ and $M_{B,y}(t)$, applied at the free end.

If the cylindrical specimen in Figure 2 undergoes bending $M_{B,x}(t)$ and torsion $M_T(t)$ moments, it is straightforward to determine the elastic peak normal and shear stresses as [30]:

$$\sigma_{z,p}(t) = \frac{32\,M_{B,x}(t)}{\pi d^3} K_{t,B}, \quad \tau_{xy,p}(t) = \frac{16\,M_T(t)}{\pi d^3} K_{t,T} \tag{4}$$

where d is the specimen diameter in the smallest section, and $K_{t,B}$ and $K_{t,T}$ are the stress concentration factors in bending and torsion. Equation (4) makes apparent how the local stress is directly proportional to the applied loadings. Moreover, it also highlights the role of the stress concentration factors, $K_{t,B}$ and $K_{t,T}$, in increasing the local stress.

Another common layout often exploited in fatigue tests is that in which the specimen is excited at its base. Figure 3 illustrates a cantilever cylindrical specimen, subjected to two orthogonal accelerations, $\ddot{y}(t)$ and $\ddot{x}(t)$, at the clamped end. Both accelerations make the specimen vibrate in bending. In most cases, though, only the vertical acceleration is applied.

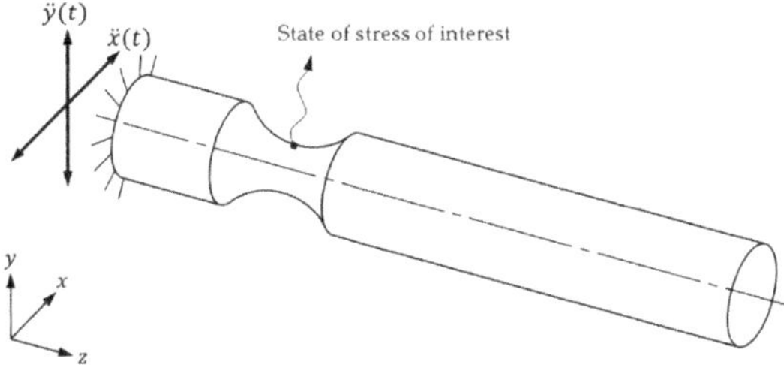

Figure 3. Notched cantilever cylindrical specimen subjected to orthogonal base accelerations $\ddot{y}(t)$ and $\ddot{x}(t)$.

In this testing layout, a modal analysis is normally carried out to identify the modes of vibration and the natural frequencies of the specimen [9]. Usually, the input acceleration is tuned at the first specimen natural frequency. Based on the harmonic analysis, the dynamic response (e.g., stress) of the system is associated to the excitation (e.g., force or acceleration); here, special care is required for kinematic excitation as the natural dynamics changes if compared to the force/dynamics excitation [9].

Assuming a time-invariant linear system, simple relationships exist between the input acceleration and the resulting local stress. For example, for a harmonic vertical acceleration $\ddot{y}(t) = a_y \cos(\omega)$ with amplitude a_y and frequency ω centered on the first resonance frequency of the specimen, the corresponding peak stress is also harmonic with same frequency and amplitude:

$$\left(\sigma_{z,p}\right)_a = |H_\sigma(\omega)| \cdot a_y \tag{5}$$

where $H_\sigma(\omega)$ is the frequency transfer function for bending. Similar relations hold for other nonzero stress components, if present. For example, an eccentric (off-set) tip mass mounted on the specimen in Figure 3 would also determine a torsional deformation of the specimen, with a corresponding shear stress at the notch. The amplitude of the peak shear stress then would be $\left(\tau_{xy,p}\right)_a = |H_\tau(\omega)| \cdot a_y$, where $H_\tau(\omega)$ is the frequency transfer function in torsion.

The previous relations can be generalized to the case of a specimen under random base accelerations:

$$S(\omega) = H(\omega)S_a(\omega)H^{*T}(\omega) \tag{6}$$

where $S_a(\omega)$ is the PSD matrix of the input accelerations, $H(\omega)$ the frequency transfer function matrix characterizing the system, and $S(\omega)$ is the PSD matrix of the stress, defined in Equation (2). The matrix $S_a(\omega)$ specifies the frequency content and correlations of the random accelerations applied to the specimen, in terms of auto- and cross-PSDs.

Before carrying out the tests, one has to carefully establish the relationship between the type and number of random loadings applied to a specimen and the resulting state of stress at the critical point. Not always does a multiaxial input determine a multiaxial state of stress. In some circumstances, a testing system under a multiaxial input develops a uniaxial stress [31,32]. An alternative to verify the state of stress evaluated theoretically or numerically (e.g., results obtained by FE analysis) is by means of strain gages. Although some specimens have a complex notched geometry that makes it difficult, if not impossible, to attach strain gages directly at the notch, they allow the local stress to be assessed indirectly through strains measured in other points of either the specimen or the testing system [17–19].

In addition to the external random loadings and the state of stress of interest, the choice of the testing machine also offers advantages and disadvantages when performing the fatigue tests; they are discussed in the next section.

3. Fatigue Testing Machines

Two different types of testing machines are usually adopted in mechanical laboratories to perform fatigue tests with random loading. They are known as servo-hydraulic, Figure 4a, and electrodynamic shaker tables, Figure 4b. Servo-hydraulic machines impose a force and/or torque as the input excitation, while electrodynamic shakers apply a force to the vibrating table (where the acceleration is usually controlled with a closed-loop control).

(a)

(b)

Figure 4. Testing machines adopted in mechanical laboratories: (**a**) servo-hydraulic; (**b**) electrodynamic shaker table.

Electrodynamic shakers have a great advantage in that they allow their table to be driven at high frequencies. Consequently, fatigue tests by shaker tables take much less time than tests by servo-hydraulic machines. On the other hand, in servo-hydraulic machines, the force or torque applied to the specimen is controlled directly by the control system; this permits the local state of stress in the specimen to be related directly to the applied loadings, see, for example, Equation (4). With electrodynamic shakers, instead, the local state of stress depends on the dynamic response of the test specimen. Carrying out a dynamic analysis to estimate the local stress in the specimen is, therefore, of considerable importance before performing fatigue tests with shaker tables.

With a uniaxial electrodynamic shaker, a specimen can easily be subjected to a random bending when excited by a vertical base acceleration; see Figure 3. However, by an appropriate setup [17–19], electrodynamic shakers can also be used for tests with bending–torsion random loadings, which result into biaxial normal and shear stresses, $\sigma_{xx}(t)$ and $\tau_{xy}(t)$. This type of bending–torsion loading is not commonly found in tests with a servo-hydraulic machine. In fact, biaxial servo-hydraulic machines can normally apply axial-torsion loading to produce a biaxial state of stress $\sigma_{xx}(t)$, $\tau_{xy}(t)$.

Servo-hydraulic machines, on the other hand, have a greater flexibility in controlling the amplitude and phase of axial-torsion loadings. In the case of fatigue testing with shaker tables, bending–torsion loadings are reached by a more complex system configuration, and their values are not controlled directly by the system. This restriction may pose some difficulty in determining the actual values of the local state of stress that is obtained in the specimen. For this reason, in applications with shaker tables, it is also important to monitor accelerations and strains. Accelerometers are often used to control the acceleration imposed on the table and the dynamic response of the system. Strain gages are employed for measuring and controlling strain at the point of maximum stress, or nearby in case it is not directly accessible. Strain gages are attached on the specimen close to a notch or hole, or on the clamping system to provide an indirect measure of the stress in the specimen.

In electrodynamic shaker tests, the fixation of the sample to the shaker armature or to the shaker table is critical; it is required that the fixation of the specimen should not have any natural frequencies in the frequency range of testing. Usually, the base acceleration of the fixation is controlled in a closed loop. Experimental transfer functions are usually based on sine-sweep, impact or random excitation and can be compared to the results obtained by using the finite element model. Proper dynamic analysis usually requires Experimental Modal Analysis (EMA) [9].

The following section makes a critical analysis of the various testing methodologies encountered in the literature, which differ not only in terms of machines but also of specimen geometries, external loads and state of stress in the critical location. Throughout the text, the terms "testing method" and "testing system" define a specific combination of testing machine, specimen geometry and load set used to perform a fatigue test.

4. Fatigue Testing Systems

Various fatigue testing systems are proposed in the literature, often with significant differences. Some systems apply deterministic (harmonic) loads, and others apply random loadings. While reviewing the systems described in the literature, this section emphasizes several important features related to fatigue testing. These features include the type of machine, specimen geometries, imposed constrains, number of input needed in terms of force and/or torque and acceleration, random loadings acting on the specimen (e.g., axial, bending, torsion, bending–torsion and tension–torsion loadings) and resulting state of stress in the critical location (e.g., uniaxial and biaxial). All these features constitute the general classification criterion adopted to classify the testing systems; see Table 1. The table groups the various systems based on common or different characteristics. It also points out whether the multiaxial state of stress applied by the system is correlated or not. Additionally, mentioned in Table 1 is the distinction of the type of specimen (i.e., plate, cylindrical and more elaborated), which is discussed in more detail later on. The classification criterion of Table 1 may represent a useful guideline to classify any new testing system.

Table 1. Classification of fatigue testing systems (C = correlated stress components; UC = uncorrelated stress components).

Machine	Specimen	Number of Inputs	Random Loading Applied to Specimen [1]					STATE OF STRESS IN Specimen			Ref.
			Ax	Be	To	Ax-To	Be-To	Uniaxial	Biaxial [2]	Biaxial [3]	
Servo-hydraulic	Plate	2	x					σ_{xx}	C		[33]
	Cylindrical	1 or 2	x		x	x		σ_{xx} or τ_{xy}		C or UC	[34]
Shaker tables	Plate	1		x				σ_{xx}			[10–14]
		1		x				σ_{xx}	C		[35–37]
	Cylindrical	1		x	x		x	σ_{xx} or τ_{xy}		C	[15]
		2		x	x		x	σ_{xx} or τ_{xy}		C or UC	[15]
		1		x			x	σ_{xx}		C	[16]
		1 or 2		x	x		x	σ_{xx} or τ_{xy}		C or UC	[17–19,38,39]
	More elaborated	1 or 2		x				σ_{xx}	UC		[20–25]

[1] Ax = axial; Be = bending; To = torsion; Ax-To = axial-torsion; Be-To = bending–torsion. [2] Biaxial with two normal stresses, σ_{xx} and σ_{yy}. [3] Biaxial with normal stress, σ_{xx}, and shear stress, τ_{xy}.

Electrodynamic shakers seem to be the most used machine, at least for the systems considered in Table 1. A possible reason for this is that the time to perform fatigue tests in high-cycle regime with random loadings is significantly shorter with shakers than servo-hydraulic machines. Another reason is that accelerated vibration tests in the automotive and aerospace industry are defined for electrodynamic shakers. Table 1 shows that the maximum number of excitations is two (e.g., vertical and horizontal accelerations), although some shakers allow three independent excitations to be applied simultaneously (e.g., vertical, horizontal and longitudinal acceleration)—in fact, only two of them are the active channels. Note also that servo-hydraulic machines in Table 1 are used to apply axial, torsion or axial-torsion loading, whereas shaker tables can apply bending, torsion or bending–torsion loadings. The type of state of stress in the specimen varies widely from one system to the other. However, cylindrical specimens can produce a pure shear stress, or combined normal and shear stresses, whereas the other specimen geometries cannot. Details of each type of specimen are described in the following sections.

4.1. Plate Specimens

Thin plate specimens with rectangular or square shape, excited at the base in vertical direction, represent the simplest and most convenient layout to produce a uniaxial state of stress in notches or holes; see [10–14] in Table 1. These systems are usually mounted on shaker tables and then excited by harmonic acceleration centered on the specimen resonant frequency. Harmonic acceleration has the advantage of easily obtaining the harmonic transfer function as the ratio between accelerations at two measurement positions. In fact, two accelerometers can be used [10–12], one attached on the base of the shaker and the second one attached at the free end of the specimen. A few accelerometers positioned along the entire length of the specimen are also observed in some applications [40,41] with the aim to obtain the modes of vibration at resonance frequencies. Although harmonic loadings allow the harmonic transfer function of the system to be determined, and they can also be used for constant amplitude fatigue tests, the random acceleration is the type of excitation most used in fatigue tests with shaker tables, so as to replicate the random loadings commonly found in engineering applications.

It is important to underline that plate specimens with rectangular or square shape, mounted on shaker tables and excited at the base, cannot produce a biaxial state of stress at critical location, but rather a uniaxial stress state. This circumstance occurs even if the specimen is excited by a multiaxial input along more than one direction, for example, vertical and horizontal accelerations, each of which can produce a bending loading. Indeed, these plate specimens of thin thickness are usually subjected to bending loadings that lead to a uniaxial state of stress in the critical location.

Plate specimens with circular shape are also employed for tests with shaker tables [35,36]. As depicted in Figure 5, such specimens are fixed at the center and then excited by a vertical random acceleration, which produces a biaxial state of stress with the critical location outside the center. The intensity of stress components can be controlled by varying the diameter or thickness of the specimen. However, the nonzero stress components are only the radial and circumferential ones. Plate specimens with a circular shape then do not develop shear stress components at the critical point; see [35,36].

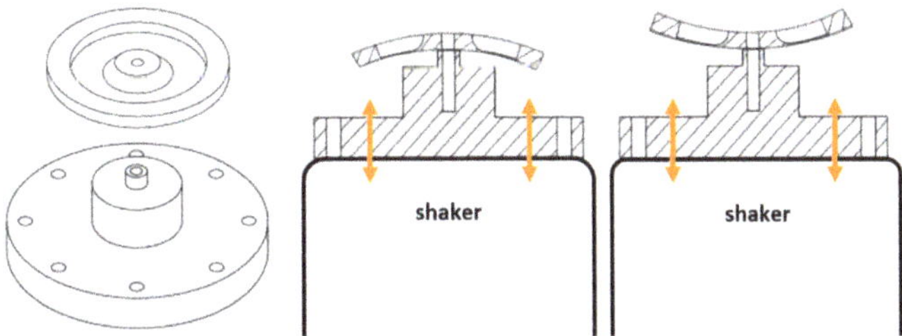

Figure 5. Testing system with a circular plate specimen excited by a uniaxial shaker (reprinted with permission from ref. [35]. Copyright 2021 Elsevier).

Plate specimens with cruciform geometry, excited by random loadings, are encountered in some fatigue tests using a servo-hydraulic machine; see [33,42] in Table 1. More often, this type of specimen is used in constant amplitude low cycle fatigue tests [43], whereas it seems not to be used in low cycle regime with random loadings.

When loaded by axial loadings applied to its two orthogonal arms (see Figure 6), the cruciform specimen can develop in the critical location a biaxial stress with two normal stress components, similarly to the circular plate specimen.

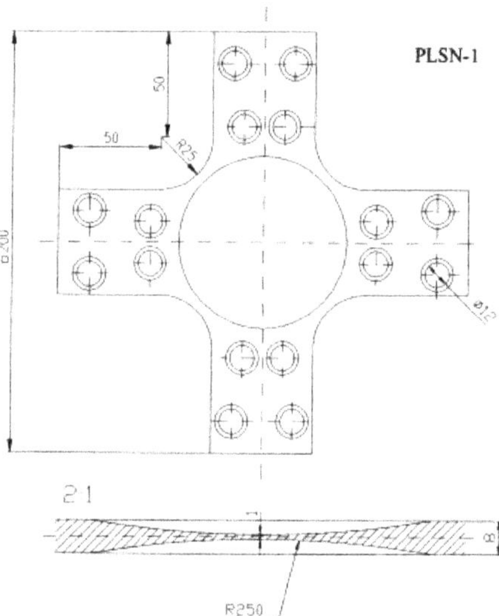

Figure 6. Plate specimens with cruciform geometry (reprinted with permission from ref. [33]. Copyright 1999 Elsevier).

Other plate specimen shapes (e.g., a particular wing shape) can be considered in tests if a biaxial stress is to be obtained by means of shaker tables; see [37] in Table 1. Though a plate specimen can develop a biaxial state of stress with two normal stresses, it cannot generate a biaxial state with normal and shear stresses. To obtain this stress, a cylindrical specimen is required.

4.2. Cylindrical Specimens

Cylindrical specimens, with or without notches, can be loaded in bending if excited by a vertical acceleration imposed by a shaker table. The acceleration can be harmonic or random, and the bending loading accordingly. In the typical layout, the specimen has one end free and the other fixed to the shaker table. The system may be instrumented by two or more accelerometers to measure accelerations at different points, and by strain gages attached on notches to control the strain [44].

In some applications, the above system is excited simultaneously by tri-axis excitations centered on specimen resonance frequencies [31,32]. Although the random loadings correspond to axial and to bending loadings in two planes, the state of stress remains uniaxial in the critical location of interest. Indeed, it is only the maximum stress position that changes, depending on the intensity and phase shift of the excitations. Therefore, the use of a tri-axial shaker does not assure that a biaxial state of stress be obtained on a cantilever cylindrical specimen. This example emphasizes that a different configuration of the test system is needed to reach a biaxial state of stress in cylindrical specimens.

To this end, a possible solution is to exploit a uniaxial shaker with a rotary table structure and a lever [15]. In one side, the cantilever cylindrical specimen is fixed to the holder structure; in the other side, it is attached to the lever. By rotating the lever arm with an arbitrary angle in the range $0 \leq \alpha \leq \pi/2$, the shaker excites the lever by imposing simultaneous bending and torsion moments to the specimen. Due to its layout, this system can only develop a coupled (correlated) bending–torsion loading, i.e., it is limited to proportional loadings. Accordingly, in Table 1, this system is listed in the fifth row (ref. [15], with one input). Choosing either of the two limit angular values, the system can apply a pure bending when the lever is parallel to the specimen axis ($\alpha = 0$), or a pure torsion when the lever is perpendicular to the specimen axis ($\alpha = \pi/2$). Strain gages are also attached on the lever to measure the strain and to control the value of normal stress in the specimen. According to the imposed angle α, it is straightforward to calculate the value of stress in any system configuration, i.e., biaxial normal and shear stresses when $0 < \alpha < \pi/2$, uniaxial pure normal stress when $\alpha = \pi/2$ and uniaxial pure shear stress when $\alpha = 0$.

By adopting a similar system layout, fatigue tests can also be performed with uncoupled (uncorrelated) bending–torsion loadings [15]. In contrast to the system described so far, now two uniaxial shakers are controlled independently. They are mounted on a table in order to excite two arms positioned perpendicularly; see Figure 7. In this case, not only coupled but also uncoupled bending–torsion random loadings can be achieved. Accordingly, in Table 1, this system is listed in the sixth row (ref. [15], with two inputs). This system yields a pure bending loading when only the arm parallel to specimen axis is excited by a uniaxial shaker. Instead, if the excitation is only imposed by the other shaker (arm perpendicular to the specimen), the specimen is subjected to torsion loading, without bending. Of course, this testing system layout needs two uniaxial shakers. Furthermore, it requires an input/output system to control the accelerations in the two shakers simultaneously.

By adding two tip masses of different weight at the free end of a cylindrical specimen excited by a uniaxial shaker table, it is possible to obtain a bending–torsion loading in the specimen; see [16] in Table 1. This layout shows that a specimen excited by a base vertical random acceleration experiences a normal and shear biaxial stress. In this layout, however, both stress components are always coupled. Their relative magnitude can be controlled by increasing or decreasing the weight ratio of the two tip masses. A pure bending loading results by selecting the same weight for both tip masses. However, obtaining the opposite loading case (only torsion) is not possible—indeed, torsion is always coupled with bending. Another aspect to mention is that this layout seems not to have been verified by experimental tests, but only by numerical simulations [16].

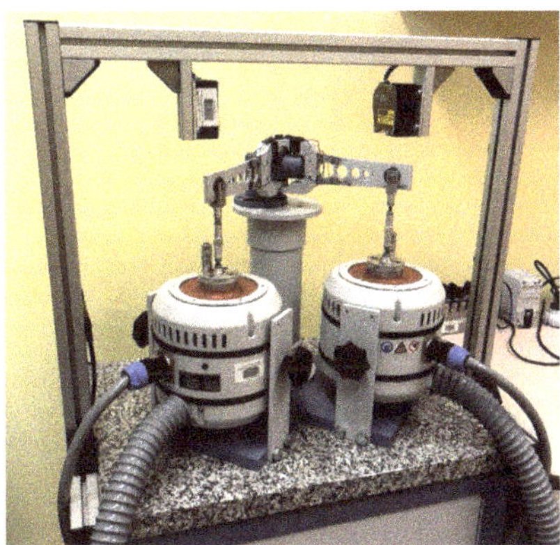

Figure 7. Testing system with a cylindrical specimen excited by two independent shakers.

A way to decouple the bending and torsion loadings—and thus improve the capabilities of the testing system in [16]—is to use a tri-axis shaker, which can apply up to three independent excitations along three orthogonal directions. Inspired by [16], a special holder structure (see Figure 8) has been designed to allow the bending and torsion random loadings to be controlled independently; see [17–19] in Table 1. In the system in Figure 8, a U-notched cylindrical specimen is fixed to a T-clamp at one end. At the other end, the specimen mounts a cantilever arm with two equal masses, and it is also constrained by a thin and flexible plate.

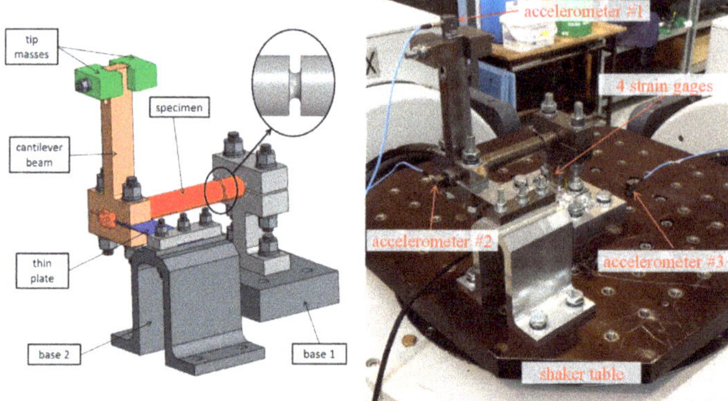

Figure 8. Testing system composed of a cylindrical specimen with eccentric tip masses, and excited by vertical and/or horizontal base accelerations (adapted with permission from ref. [19]. Copyright 2018 Elsevier).

This thin and flexible plate constraints any horizontal movement but allows the rotation of the specimen end. The plate thus prevents the specimen from being subjected to bending in horizontal direction, but it is very thin to allow the specimen to rotate around its axis when subjected to torsion. Therefore, when the specimen is excited by a vertical base acceleration, Figure 9a, it vibrates in the vertical plane (only bending loading). When, instead, it is excited by a horizontal base acceleration, Figure 9b, it vibrates in the horizontal plane (only torsion loading caused by the oscillation of the two eccentric masses).

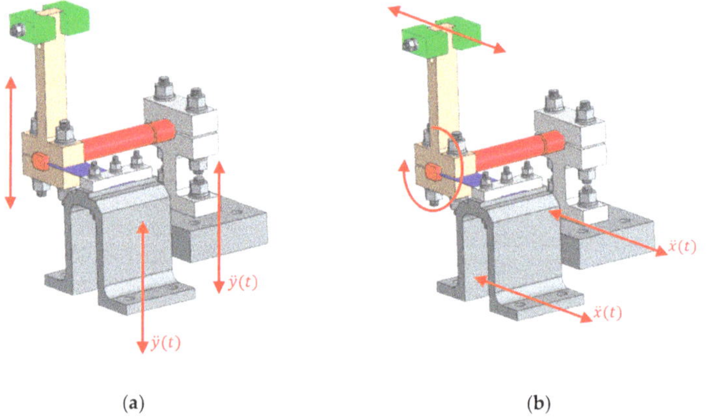

Figure 9. Testing system with a cylindrical specimen excited by (**a**) vertical base acceleration and (**b**) horizontal base acceleration.

The testing system has accelerometers to monitor the accelerations of the shaker table in closed-loop control, and the acceleration of the extremity of the specimen and cantilever arm. As it is not possible to use strain gages to measure the strains directly at the specimen notch (being it too small), an indirect measure it performed. Strain gages are indeed glued onto the lateral faces of the T-clamp, so as to provide a measure of the bending moments there and, indirectly, of the bending and torsion loadings and, accordingly, of the resulting normal and shear stress at the specimen notch.

Note that the tri-axis shaker cannot excite only one single axis or two axes at a time, keeping the others at rest. All three axes need to be active simultaneously, which appears to be a small limitation if tests with two, or even only one, accelerations need to be carried out. However, this circumstance can easily be overcome by setting a very low level of acceleration on the "secondary" axes that, theoretically, should not be activated.

It is finally worth to mention that the system layout described so far, as shown in Figure 8, has been adopted in [38,39], though with two lateral thin plates instead of one, to perform bending–torsion fatigue tests.

Hollow cylindrical specimens subjected to axial, torsion and internal pressure perhaps represent the most versatile testing system in terms of the state of stress achievable. Indeed, this system allows a three-dimensional stress state (two normal stresses and one shear stress) to be obtained by servo-hydraulic machines and pressure chambers [45,46]. On the other hand, it may be presumed that loading a hollow specimen in axial, torsion and simultaneously with an internal pressure, requires special-purpose equipment that is likely to be more expensive than the usual testing machines found in laboratories.

A hollow cylindrical specimen with a small hole perpendicular to its axis, loaded by a servo-hydraulic machine, may be subjected to an uncoupled biaxial state of stress; see [34] in Table 1. This specimen is fixed at both ends, where the servo-hydraulic machine applies an axial-torsion random loading. The biaxial state of stress at the hole can be monitored with minimal difficulty by controlling the input force and/or torque and relating it to the corresponding stress components. Once again, this system configuration emphasizes how servo-hydraulic machines offer a greater simplicity over shaker tables in the direct control of both the intensity and phase shift between axial-torsion loading actions on cylindrical specimens. Instead, cantilever cylindrical specimens excited by shakers require the dynamic response of the specimen fixed on the holder system to be determined in order to evaluate the normal and shear stress values at the critical location.

4.3. More Elaborated Specimens

More elaborated specimens are developed with the aim to perform accelerated fatigue tests using shaker tables and simulate a real-world scenario of complex structures. A Y-shaped specimen with a central hole and two masses at the free ends has been proposed; see Figure 10. The Y-shaped specimen is made by three rectangular cross-sections arranged at 120° around the hole and has in the frequency range up to 2 kHz several natural frequencies that can be vibration fatigued. The attached masses can be used to adjust the frequencies of particular natural frequencies [21]. In [22], it was shown that the internal damping has a significant influence on the fatigue damage. In [20], multiaxial loads were achieved by exciting two mode shapes (one in the vertical and one in the horizontal direction). In vibration fatigue, the specimen is typically considered as broken when the natural frequency drops by 2–5% (different values are used in different studies); see e.g., [9]. Fatigue parameters (fatigue strength and inverse slope of S–N curve) need to be experimentally identified using the specimen; see e.g., [23], where 10 specimens were used. It may finally be presumed that manufacturing several specimens with more elaborated geometry has a cost slightly higher than producing specimens with simpler shapes.

Figure 10. Testing system including Y-shaped specimen with a central hole and two masses at the free ends subjected to horizontal force and vertical acceleration (reprinted with permission from ref. [23]. Copyright 2016 Elsevier).

In [20], two electrodynamic shakers were used: one for vertical random excitation (for the excitation of first natural frequency) and one for horizontal excitation (for the second natural frequency). Masses attached to the Y-sample allowed modal frequencies to be adjusted to the needs. The test system is instrumented with accelerometers positioned at different points and a strain gage attached in the critical region. The accelerometers are used to monitor in real time the dynamic response of system, which in turn updates the FE model. The vertical excitations are controlled in a closed loop with measurements of accelerometers. Another excitation is applied perpendicular to the shaker vertical axis. It is imposed close to the specimen hole and monitored by a force transducer.

This special Y-shaped system is excited by two uncoupled random loadings (along vertical and horizontal directions); see [20–25] in Table 1. The system can develop an uncoupled biaxial state of stress at the critical location in terms of normal stress components.

5. Conclusions

This paper presented an overview of the various fatigue testing systems used for subjecting metals to random loadings, as they are described in relevant articles from the literature. The presented overview of relevant works compared the different testing systems in terms of several characteristics, namely, the type of machine (e.g., servo-hydraulic and shaker tables), specimen geometry, number of inputs needed to carry out a fatigue test, random loadings acting on the specimen and resulting local state of stress. For each of the above characteristics, Table 2 summarizes the specific features adopted for classifying the various systems. Based on the specific features in Table 2, all the analyzed testing systems were also classified into a more comprehensive table (named Table 1 in the text), which allowed the analogies, differences, advantages and possible limitations to be emphasized in a clear way. Both tables also summarize the main criteria that may be used, in the future, as a guideline to classify new equipment.

Table 2. Characteristics and specific features used to classify fatigue testing systems.

Machine	Specimen Geometry	Random Loading [1]	State of Stress
Servo-hydraulic orshaker table	Plate, cylindrical or more elaborated	Ax, Be, To, Ax-To or Be-To	Uniaxial, biaxial with σ_{xx} and σ_{yy} or biaxial with σ_{xx} and τ_{ry}

[1] Ax = axial; Be = bending; To = torsion; Ax-To = axial-torsion; Be-To = bending–torsion.

Author Contributions: Writing—original draft preparation, J.M.E.M. and D.B.; writing—review and editing, J.M.E.M., D.B., A.N. and J.S.; supervision, D.B., A.N. and J.S. All authors have read and agreed to the published version of the manuscript.

Funding: This research received no external funding.

Conflicts of Interest: The authors declare no conflict of interest.

References

1. Benasciutti, D. *Fatigue Analysis of Random Loadings: A Frequency Domain Approach*; LAP Lambert Academic Publishing: Saarbrücken, Germany, 2012; ISBN 9783659123702.
2. Benasciutti, D.; Sherratt, F.; Cristofori, A. Recent developments in frequency domain multi-axial fatigue analysis. *Int. J. Fatigue* **2016**, *91*, 397–413. [CrossRef]
3. Du, J.; Li, H.; Zhang, M.; Wang, S. A novel hybrid frequency-time domain method for the fatigue damage assessment of offshore structures. *Ocean Eng.* **2015**, *98*, 57–65. [CrossRef]
4. Li, Z.; Ringsberg, J.W.; Storhaug, G. Time-domain fatigue assessment of ship side-shell structures. *Int. J. Fatigue* **2013**, *55*, 276–290. [CrossRef]
5. Kocabicak, U.; Firat, M. A simple approach for multiaxial fatigue damage prediction based on FEM post-processing. *Mater. Des.* **2004**, *25*, 73–82. [CrossRef]
6. Marques, J.M.E.; Benasciutti, D.; Carpinteri, A.; Spagnoli, A. An algorithm for fast critical plane search in computer-aided engineering durability analysis under multiaxial random loadings: Application to the Carpinteri–Spagnoli–Vantadori spectral method. *Fatigue Fract. Eng. Mater. Struct.* **2020**, *43*, 1978–1993. [CrossRef]
7. Benasciutti, D.; Marques, J.M.E. An efficient procedure to speed up critical plane search in multiaxial fatigue: Application to the Carpinteri-Spagnoli spectral criterion. *MATEC Web Conf.* **2019**, *300*, 16003. [CrossRef]
8. Mršnik, M.; Slavič, J.; Boltežar, M. Vibration fatigue using modal decomposition. *Mech. Syst. Signal Process.* **2018**, *98*, 548–556. [CrossRef]
9. Slavič, J.; Boltežar, M.; Mršnik, M.; Česnik, M.; Javh, J. *Vibration Fatigue by Spectral Methods: From Structural Dynamics to Fatigue Damage-Theory and Experiments*, 1st ed.; Elsevier: Amsterdam, The Netherlands, 2020; ISBN 9780128221907.
10. Khalij, L.; Gautrelet, C.; Guillet, A. Fatigue curves of a low carbon steel obtained from vibration experiments with an electrodynamic shaker. *Mater. Des.* **2015**, *86*, 640–648. [CrossRef]
11. Gautrelet, C.; Khalij, L.; Appert, A.; Serra, R. Linearity investigation from a vibratory fatigue bench. *Mech. Ind.* **2019**, *20*, 101. [CrossRef]
12. Appert, A.; Gautrelet, C.; Khalij, L.; Troian, R. Development of a test bench for vibratory fatigue experiments of a cantilever beam with an electrodynamic shaker. *MATEC Web Conf.* **2018**, *165*, 10007. [CrossRef]
13. Ellyson, B.; Brochu, M.; Brochu, M. Characterization of bending vibration fatigue of SLM fabricated Ti-6Al-4V. *Int. J. Fatigue* **2017**, *99*, 25–34. [CrossRef]
14. Ghielmetti, C.; Ghelichi, R.; Guagliano, M.; Ripamonti, F.; Vezzù, S. Development of a fatigue test machine for high frequency applications. *Procedia Eng.* **2011**, *10*, 2892–2897. [CrossRef]
15. Łagoda, T.; Macha, E.; Niesłony, A. Fatigue life calculation by means of the cycle counting and spectral methods under multiaxial random loading. *Fatigue Fract. Eng. Mater. Struct.* **2005**, *28*, 409–420. [CrossRef]
16. Nguyen, N.; Bacher-Höchst, M.; Sonsino, C.M. A frequency domain approach for estimating multiaxial random fatigue life. *Mater. Werkst.* **2011**, *42*, 904–908. [CrossRef]
17. Zanellati, D.; Benasciutti, D.; Tovo, R. Vibration fatigue tests by tri-axis shaker: Design of an innovative system for uncoupled bending/torsion loading. *Procedia Struct. Integr.* **2018**, *8*, 92–101. [CrossRef]
18. Zanellati, D.; Benasciutti, D.; Tovo, R. Fatigue strength of S355JC steel under harmonic and random bending-torsion loading by a tri-axis shaker: Preliminary experimental results. *MATEC Web Conf.* **2019**, *300*, 17006. [CrossRef]
19. Zanellati, D.; Benasciutti, D.; Tovo, R. An innovative system for uncoupled bending/torsion tests by tri-axis shaker: Numerical simulations and experimental results. *MATEC Web Conf.* **2018**, *165*, 16006. [CrossRef]
20. Česnik, M.; Slavič, J.; Čermelj, P.; Boltežar, M. Frequency-based structural modification for the case of base excitation. *J. Sound Vib.* **2013**, *332*, 5029–5039. [CrossRef]
21. Ogrinec, P.; Slavič, J.; Česnik, M.; Boltežar, M. Vibration fatigue at half-sine impulse excitation in the time and frequency domains. *Int. J. Fatigue* **2019**, *123*, 308–317. [CrossRef]
22. Palmieri, M.; Česnik, M.; Slavič, J.; Cianetti, F.; Boltežar, M. Non-Gaussianity and non-stationarity in vibration fatigue. *Int. J. Fatigue* **2017**, *97*, 9–19. [CrossRef]
23. Mršnik, M.; Slavič, J.; Boltežar, M. Multiaxial vibration fatigue—A theoretical and experimental comparison. *Mech. Syst. Signal Process.* **2016**, *76–77*, 409–423. [CrossRef]
24. Mršnik, M.; Slavič, J.; Boltežar, M. Multiaxial Fatigue Criteria for Random Stress Response–Theoretical and Experimental Comparison. *Procedia Eng.* **2015**, *101*, 459–466. [CrossRef]
25. Česnik, M.; Slavič, J.; Boltežar, M. Uninterrupted and accelerated vibrational fatigue testing with simultaneous monitoring of the natural frequency and damping. *J. Sound Vib.* **2012**, *331*, 5370–5382. [CrossRef]
26. Susmel, L. *Multiaxial Notch Fatigue*, 1st ed.; Woodhead Publishing: Cambridge, UK, 2009; ISBN 9781845695828.
27. Morettini, G.; Braccesi, C.; Cianetti, F.; Razavi, S.M.J.; Solberg, K.; Capponi, L. Collection of experimental data for multiaxial fatigue criteria verification. *Fatigue Fract. Eng. Mater. Struct.* **2019**, *43*, 162–174. [CrossRef]
28. Socie, D.; Marquis, G. *Multiaxial Fatigue*; SAE International: Warrendale, PA, USA, 2000; ISBN 9780768065107.

29. Benasciutti, D.; Zanellati, D.; Cristofori, A. The "Projection-by-Projection" (PbP) criterion for multiaxial random fatigue loadings. *Frattura Integrità Strutt.* **2018**, *13*, 348–366. [CrossRef]
30. Pilkey, W.D.; Pilkey, D.F. *Peterson's Stress Concentration Factors*, 3rd ed.; John Wiley & Sons: Hoboken, NJ, USA, 2008; ISBN 978047004824.
31. Luo, Z.; Chen, H.; He, X.; Zheng, R. Two time domain models for fatigue life prediction under multiaxial random vibrations. *Proc. Inst. Mech. Eng. Part C J. Mech. Eng. Sci.* **2019**, *233*, 4707–4718. [CrossRef]
32. Whiteman, W.E.; Berman, M.S. Fatigue failure results for multi-axial versus uniaxial stress screen vibration testing. *Shock Vib.* **2002**, *9*, 319–328. [CrossRef]
33. Łagoda, T.; Macha, E.; Bedkowski, W. A critical plane approach based on energy concepts: Application to biaxial random tension-compression high-cycle fatigue regime. *Int. J. Fatigue* **1999**, *21*, 431–443. [CrossRef]
34. Niesłony, A.; Růžička, M.; Papuga, J.; Hodr, A.; Balda, M.; Svoboda, J. Fatigue life prediction for broad-band multiaxial loading with various PSD curve shapes. *Int. J. Fatigue* **2012**, *44*, 74–88. [CrossRef]
35. Morettini, G.; Braccesi, C.; Cianetti, F.; Razavi, S. Design and implementation of new experimental multiaxial random fatigue tests on astm-a105 circular specimens. *Int. J. Fatigue* **2021**, *142*, 105983. [CrossRef]
36. Morettini, G.; Braccesi, C.; Cianetti, F. Experimental multiaxial fatigue tests realized with newly developed geometry specimens. *Fatigue Fract. Eng. Mater. Struct.* **2018**, *42*, 827–837. [CrossRef]
37. George, T.J.; Seidt, J.; Shen, M.-H.H.; Nicholas, T.; Cross, C.J. Development of a novel vibration-based fatigue testing methodology. *Int. J. Fatigue* **2004**, *26*, 477–486. [CrossRef]
38. Luo, Z.; Chen, H.; He, X. Influences of correlations between biaxial random vibrations on the fatigue lives of notched metallic specimens. *Int. J. Fatigue* **2020**, *139*, 105730. [CrossRef]
39. Luo, Z.; Chen, H.; Zheng, R.; Zheng, W. A damage gradient model for fatigue life prediction of notched metallic structures under multiaxial random vibrations. *Fatigue Fract. Eng. Mater. Struct.* **2020**, *43*, 2101–2115. [CrossRef]
40. Kim, C.-J.; Lee, B.-H.; Kang, Y.J.; Ahn, H.-J. Accuracy enhancement of fatigue damage counting using design sensitivity analysis. *J. Sound Vib.* **2014**, *333*, 2971–2982. [CrossRef]
41. Kim, C.-J.; Kang, Y.J.; Lee, B.-H.; Ahn, H.-J. Design sensitivity analysis of a system under intact conditions using measured response data. *J. Sound Vib.* **2012**, *331*, 3213–3226. [CrossRef]
42. Aid, A.; Bendouba, M.; Aminallah, L.; Amrouche, A.; Benseddiq, N.; Benguediab, M. An equivalent stress process for fatigue life estimation under multiaxial loadings based on a new non linear damage model. *Mater. Sci. Eng. A* **2012**, *538*, 20–27. [CrossRef]
43. Itoh, T.; Sakane, M.; Ohnami, M. High temperature multiaxial low cycle fatigue of cruciform specimen. *J. Eng. Mater. Technol.* **1994**, *116*, 90–98. [CrossRef]
44. Kim, C.-J.; Kang, Y.J.; Lee, B.-H. Experimental spectral damage prediction of a linear elastic system using acceleration response. *Mech. Syst. Signal. Process.* **2011**, *25*, 2538–2548. [CrossRef]
45. Papuga, J.; Fojtík, F. Multiaxial fatigue strength of common structural steel and the response of some estimation methods. *Int. J. Fatigue* **2017**, *104*, 27–42. [CrossRef]
46. Fojtík, F.; Papuga, J.; Fusek, M.; Halama, R. Validation of multiaxial fatigue strength criteria on specimens from structural steel in the high-cycle fatigue region. *Materials* **2020**, *14*, 116. [CrossRef] [PubMed]

Article

Numerical and Experimental Investigation of Cumulative Fatigue Damage under Random Dynamic Cyclic Loads of Lattice Structures Manufactured by Laser Powder Bed Fusion

Marco Pisati, Marco Giuseppe Corneo, Stefano Beretta, Emanuele Riva, Francesco Braghin and Stefano Foletti *

Department of Mechanical Engineering, Politecnico di Milano, 20156 Milan, Italy; marco.pisati@polimi.it (M.P.); giuseppemarco.corneo@mail.polimi.it (M.G.C.); stefano.beretta@polimi.it (S.B.); emanuele.riva@polimi.it (E.R.); francesco.braghin@polimi.it (F.B.)
* Correspondence: stefano.foletti@polimi.it

Citation: Pisati, M.; Corneo, M.G.; Beretta, S.; Riva, E.; Braghin, F.; Foletti, S. Numerical and Experimental Investigation of Cumulative Fatigue Damage under Random Dynamic Cyclic Loads of Lattice Structures Manufactured by Laser Powder Bed Fusion. *Metals* **2021**, *11*, 1395. https://doi.org/10.3390/met11091395

Academic Editor: Christian Mittelstedt

Received: 28 July 2021
Accepted: 31 August 2021
Published: 3 September 2021

Publisher's Note: MDPI stays neutral with regard to jurisdictional claims in published maps and institutional affiliations.

Abstract: Lattice structures are lightweight engineering components suitable for a great variety of applications, including those in which the structural integrity under vibration fatigue is of paramount importance. In this work, we experimentally and numerically investigate the dynamic response of two distinct lattice configurations, in terms of fatigue damage and life. Specifically, Face-Centered-Cubic (FCC) and Diamond lattice-based structures are numerically studied and experimentally tested under resonant conditions and random vibrations, until their failure. To this end, Finite Element (FE) models are employed to match the dynamic behavior of the system in the neighborhood of the first natural frequency. The FE models are employed to estimate the structural integrity by way of frequency and tip acceleration drops, which allow for the identification of the failure time and a corresponding number of cycles to failure. Fatigue life under resonant conditions is well predicted by the application of conventional multiaxial high cycle fatigue criteria to the local state of stress. The same approach, combined with the Rainflow algorithm and Miner's rule, provides good results in predicting fatigue damage under random vibrations.

Keywords: lattice structures; structural dynamic response; vibration fatigue testing; fatigue life prediction

1. Introduction

The exploitation of periodic lattice structures has lately biased the way lightweight components are designed, which is motivated by a number of static and dynamic properties that are not commonly achievable with conventional materials and configurations, such as high strength to weight ratio [1], negative Poisson's ratio [2], improved damping capabilities [3,4], among others. Relevant examples include lattices characterized by peculiar topologies, see Ref. [5] for a review of the state of the art. Such configurations have sparked broad interest within several realms of engineering, such as aerospace, maritime engineering, automotive and biomedical. For instance, Lira et al. [6] employed a lattice structure to reduce the inertial forces of a fan blade by way of a graded 2D honeycomb core. Spadoni et al. [7] have functionally designed chiral-based auxetic cells to improve the response of airfoils in terms of dissipation.

In this context, additive manufacturing (AM) techniques are often used for a large variety of applications and suit the fabrication of many lattice structures present in the literature. Indeed, the continuous technological improvement of AM techniques has led to the fabrication of complex structures impossible to obtain with conventional manufacturing processes. Relevant examples include 3D printing techniques. For example, Colombo et al. [8] observed that the damping of AlSi10Mg manufactured by Selective Laser Melting (SLM) can be strongly affected by the printing directions. Fiocchi et al. [9,10]

compared the damping of a lattice and full solid structures made of SLM Ti6Al4V, observing a two-orders of magnitude increase in the case of lattice structures. Similar results have been observed in a number of papers [7,11], which further confirms the improved dynamic performances of lattice structures. Practically, industrial applications require reliable methods capable of predicting the structural integrity of such components. A number of static and fatigue criteria have been presented in recent works, see Ref. [12] for a recent review of the state of the art on this matter.

Several experimental studies have been proposed [13–16] in the attempt to shed light on the fatigue behavior of a variety of lattice structures. The effect of porosity on the fatigue strength of cellular materials manufactured by SLM has been investigated in Ref. [17]. Different structures with porosities between 68% and 84% have been manufactured and dynamically tested under fatigue. It is shown that all fatigue data points collapse on a single Stress-Life (SN) curve when the applied stress amplitude is normalized with respect to the yield strength of the structures. However, it is also found that the normalized endurance limit of all tested porous structures is lower than that of the corresponding solid material, i.e., 10–20% of the yield strength of the lattice structure against 40% for the bulk material. This discrepancy has been attributed to the transitions in cell geometry and to defects induced during the printing process. Even if the cell topology is optimized for reducing the stress concentration in the transitions [18], the effect of the printing process on the surface roughness and on a non-uniform thickness of the struts can reduce fatigue life and strength.

Several studies have been proposed in the attempt to reduce the number of defects induced by the printing process of L-PBF. The effects of laser power, scanning speed, beam diameter, and powder variability on metal components have been studied in Refs. [19–23]. Regarding the surface roughness, a significant improvement has been offered in [24] through Laser polishing (LP). Despite the continuous technological improvement of L-PBF, the manufacturing of lattice structure still remains quite challenging due to the small feature size and complex shape. In this regard, Dallago et al. [25] observed that node geometry and printing directions have a relevant impact on the fatigue behavior of lattice structures and can be tailored to improve fatigue life. In addition, the dependence from defects and strut irregularities has been addressed numerically by Boniotti et al. [26] by way of a finite element model based on the as-manufactured geometry, which has been reconstructed by CT-scan. Then, a multiaxial high cycle fatigue criterion has been used to prove that the highest equivalent stress occurs in correspondence of the strut nodes and on the largest strut irregularities.

Better fatigue performance has been observed for triply periodic minimal surface (TPMS) structures, where the absence of struts and nodes significantly reduces the stress concentrations, and a continuous curvature of their surface provides great benefits in terms of fatigue strength [27–29]. In other words, there is a large variety of studies about static and fatigue criteria of lattice structures. However, the dynamic characterization under resonance and the random fatigue behavior has been elusive from the practical perspective and still represents an open point.

In this context, the fatigue assessment of porous structures has been so far performed only experimentally through imposed random accelerations [30]. In detail, Richard et al. developed an innovative loop heat pipe for a Nasa's cubesat, with a porous wick fabricated by AM techniques, able to resist to a given power spectral density (PSD), which is representative of the acceleration of the satellite launch phase. In the proposed configuration, the structural integrity has been verified by way of an experimental campaign. However, for design purposes, a numerical method to predict fatigue life under random loads is often preferable, but existing vibration fatigue criteria have not been verified for lattice structures yet.

Motivated by this lack in the literature, in this work we present and experimentally validate a method for fatigue life prediction for face-centered-cubic (FCC) and Diamond lattice structures, manufactured by Laser Powder Bed Fusion (L-PBF), under resonant condition and random vibrations.

Two distinct types of experimental tests have been performed to study the fatigue behavior: (i) an endurance test imposing an excitation frequency corresponds to the first bending mode, (ii) a random excitation fatigue test to meet excitation levels and spectra consistent with real applications. Before fatigue tests, the frequency response function of all specimens, including all the relevant dynamic quantities such as damping and natural frequencies, has been estimated through sweep frequency tests. A numerical model with induced localized damage, broken strut, or missing node, is employed to emulate the effect of a local failure of the lattice elements to the drops in the global dynamic response in resonance and due to random input. The failure condition is defined as corresponding to the failure of the most loaded strut in the numerical analysis.

Fatigue test results under resonant conditions, i.e., at constant stress amplitude, have been analyzed through the procedure presented in Ref. [26]; i.e., by computing an equivalent stress amplitude based on a multiaxial high cycle fatigue criterion. In addition, a map of the equivalent stress amplitude has been obtained through finite element analyses for different values of tip displacement and damping ratios. By using the equivalent stress amplitude value in the SN diagram of the parent material, corresponding to the tip displacement and the damping ratio experimentally measured in the endurance tests, fatigue life can be predicted with good accuracy.

For the experimental tests under random vibrations, starting from the application of the Rainflow's algorithm to the experimentally measured tip displacement, the cumulative fatigue damage has been computed with the Miner's rule where, for every cycle, the damage is calculated considering the corresponding tip displacement, i.e., the equivalent stress amplitude thorough the stress map, and the SN diagram of the parent material.

The proposed procedure leads to fatigue damage to failure, which is around 1, demonstrating an accurate matching between the predictive model and experimental results.

2. Data and Experimental Methods

This section includes the geometrical and physical data of the lattice structures employed in this work and summarizes the experimental setup employed to characterize and study the fatigue behavior of the lattice structures.

2.1. Material, Cell Topology and Specimen Shape and Dimension

The bulk material used for the printing process of the FCC and Diamond lattices is the AlSi10Mg, whose mechanical properties are reported in Table 1. Two different periodic truss structures are selected and printed by L-PBF. For the first one, the periodic structure is obtained by the combination of 8 Simple Cubic and 1 Face Centered Cubic (FCC) cell. For the second one, a Diamond cell topology is used. Both configurations are designed with the unit length and strut diameter of 5 and 0.6 mm, respectively, while the corresponding relative density (ρ) is 7.3% and 7.0% for FCC and Diamond cell, respectively.

Table 1. Mechanical properties for the bulk material used for specimen fabrication.

Material	Elastic Modulus [MPa]	Shear Modulus [MPa]	Poisson's Ratio	Yield Stress [MPa]
AlSi10Mg	68,000	26,154	0.3	240

This specimen design is meant to be representative of a sandwich panel used as a heat dissipator with an upper and lower solid skin of 1.5 mm, see Figure 1. The core is characterized by two unit cells along the vertical direction, for a total height of the specimen of 13 mm, while four unit cells are considered along the lateral direction, corresponding to a total width of 20 mm. Such a configuration is designed to investigate the structural behavior under bending vibration, and, for this reason, the longitudinal dimension is designed to be dominant. The specimen length with 19 unit cells, for a total length of 95 mm, leads to a lattice gauge section of $4 \times 2 \times 19$. Due to experimental requirements, the specimens are equipped with a solid part that serves as a rigid connection between the lattice structure and the shaker. Such a region has a circular geometry with a diameter of 42 mm with six equally spaced holes for proper clamping on the shaker. A transition region characterized by a growing lattice section (C-shaped fillet) is designed between the clamped solid and lattice structure to avoid possible stress concentration and fatigue damage near the clamped region in the solid-lattice interface. All the specimens were printed with a SLM280-1.0 printer (SLM-Solutions, Lubeck, Germany) driven with laser power of 350 W, scanning speed of 1150 mm/s, layer thickness of 50 μm, and hatching distance of 0.17 mm. The angle between the longitudinal direction of the sample and the build direction was set equal to $30°$. Such process parameters have been selected as the result of an optimization procedure for printing lattice structures.

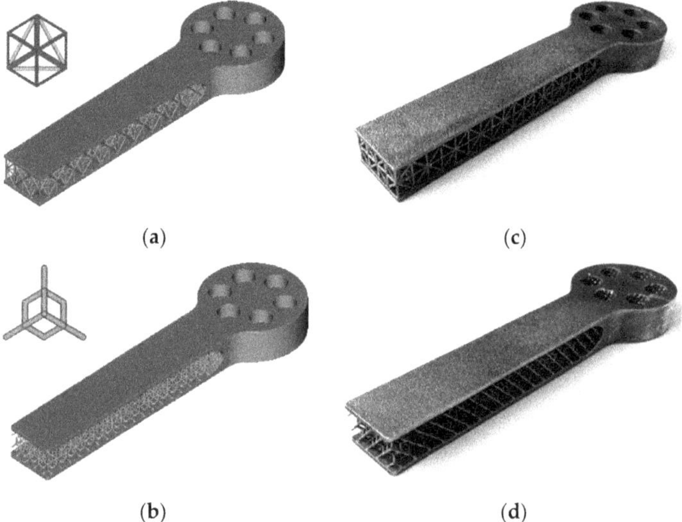

Figure 1. (**a,b**) FCC and Diamond specimen modelled with nTopology. (**c,d**) FCC and Diamond lattice specimens made by L-PBF.

2.2. Experimental Tests and Setup

Three distinct types of experimental tests have been performed on the lattice specimens in the attempt to study their fatigue behavior: (i) a linear sweep test is firstly carried out for the dynamic characterization of the structure; (ii) an endurance and (iii) a random test have been designed to study the fatigue response.

For all the tests, the lattice sandwich is excited through an imposed vertical acceleration, which is feedback-controlled for the entire duration of the experiments. To this end, an air-cooled S452 series (Unholtz-Dickie Corp., Wallingford, CT, USA) one-axis electrodynamic shaker (as shown in Figure 2a) is employed, which can provide a maximum acceleration and force of 90 g and 20 KN. The acceleration control is implemented using a ± 50 g PCB accelerometer (with a sensitivity of 104 mV/g) rigidly connected to the shaker piston and able to measure the vertical component of the acceleration. Two different accelerometers have been bonded to measure the base and tip accelerations. A triaxial accelerometer with a measurement range of ± 500 g and a sensitivity of 10.41 mV/g is used to measure the tip acceleration, while two PCB piezoelectric uniaxial accelerometers with a measurement range ± 50 g and a sensitivity of 100.2 mV/g have been bonded on the base of the specimen. Such a measurement point corresponds to the input location in the numerical models, as will be discussed more in detail later in the paper. The accelerometers are connected to a Sensor Signal Control Model 483C Series (PCB Piezotronic Inc., Depew, NY, USA) which, in turn, is connected to the data acquisition system (DAQ), consisting of a NI 9239 board, able to save the measured data with a sampling frequency of 5 kHz.

(a)　　　　　　　　　　　　　　　　(b)

Figure 2. (**a**) Lateral view of the electrodynamic shaker with mounted FCC specimen. (**b**) Axonometric view of the structure. A set of accelerometers is bonded on the tip, base and shaker piston, while a pair of masses have been rigidly connected on the tip to tailor the desired dynamic behavior.

Under this configuration, the imposed acceleration can be considered sufficiently accurate and stable enough to perform the tests, provided that the tightening torque does not affect the dynamic behavior of the specimen. Finally, a pair of masses are placed at the tip of the specimen in such a way to preserve a symmetric mass distribution when the accelerometer is connected to the structure, as shown in Figure 2b. The total added mass is 50 g.

2.2.1. Sweep Frequency Test

The aim of the sweep test is to identify the first natural frequency, damping ratio and Frequency Response Function (FRF) of the FCC and Diamond lattice specimens. Such information is necessary to qualify the real stresses induced within the lattice structure and to validate the numerical model, which is preparatory for the fatigue life estimation. Practically, the frequency sweep is provided in terms of constant amplitude acceleration spanning a defined range of frequencies, which are centered in the neighborhood of the first natural frequency for both the FCC and Diamond configurations. That is, between 375–425 Hz and 325–375 Hz respectively for a time of 200 s. The acceleration levels for FCC and Diamond lattice specimen are set equal to 0.2 g and 0.11 g respectively, which is sufficiently low to avoid nonlinear interactions or failures within the lattice. The FRF is evaluated using the H1 estimator considering the base and tip accelerations as input and output. For the results hereafter presented, the goodness of the estimation is verified by way of the coherence function γ, which is observed to be always close to 1. The first natural frequency ω_0 and corresponding damping ratio ζ are estimated through the frequency response function by way of the Half-Power Method. That is:

$$\zeta = \frac{\omega_2^2 - \omega_1^2}{4\omega_0^2}$$

which is considered independent of the amplitude at this stage of the work.

2.2.2. Endurance Test

The endurance tests have been performed imposing a monochromatic excitation. This is necessary to define the SN curve for the fatigue life of the lattice prototypes. The excitation frequency corresponds to the first bending mode of the lattice sandwich. The state of stress of the prototype is estimated combining the measured acceleration at the tip and at the clamp and the numerical model. In the case at hand, five FCC and four Diamond specimens, have been tested in resonance imposing a base acceleration necessary to achieve tip displacements of 0.31 and 0.2 mm. From the numerical models, the corresponding stresses are estimated, as will be discussed in the following sections. At failure of some of the lattice elements, there is a change in the natural frequency of the prototype. After this point, the lattice structure experiences a softening behavior that is more pronounced for damaged specimens. For the scope of this work, a linear model and linear identification of the relevant parameters is considered sufficiently accurate to capture the first-order variations of the natural frequency, which is a blueprint of the local degradation within the core of the lattice sandwich.

2.2.3. Random Test

Random excitation fatigue tests are also considered to meet excitation levels and spectra that are more consistent with real applications, especially in the context of aerospace and mechanical engineering. To this end, the PSD reported in Table 2 has been employed to investigate the fatigue life.

Table 2. PSD of imposed acceleration.

Frequency [Hz]	Amplitude [g^2/Hz]
15	0.04
100	0.04
300	0.17
1000	0.17
2000	0.05

Such a spectrum is scaled to meet two distinct excitation levels (the data in Table 2 are reduced by a factor of 10 and 15, respectively) and are applied for 1 and 2 h in the same experimental configuration detailed in Section 2.2. It is expected that during the experimental tests, the specimen experiences a modification of the first natural frequency which can be captured by a careful post-processing of the acceleration levels.

3. Numerical Models

The numerical models employed in this manuscript are of fundamental importance to provide an estimate of the stress levels within the lattice structure starting from the accelerations measured during the experimental tests. To this end, four distinct numerical models have been developed using the commercial software Abaqus. Specifically, a discrete model and a homogeneous one have been considered for each specimen type. The discrete model describes the full three-dimensional topology and dynamics of the FCC and Diamond configurations, while in the homogeneous models the lattice structure is replaced by a solid with equivalent properties. To simulate the homogenized response of the lattice structures, single cell models were developed and Periodic Boundary Conditions (PBC) were implemented coupling the degrees of freedom of different points on opposite faces of the reference volume element with a proper set of equations [31,32]. A graphical representation of such models is illustrated in Figure 3.

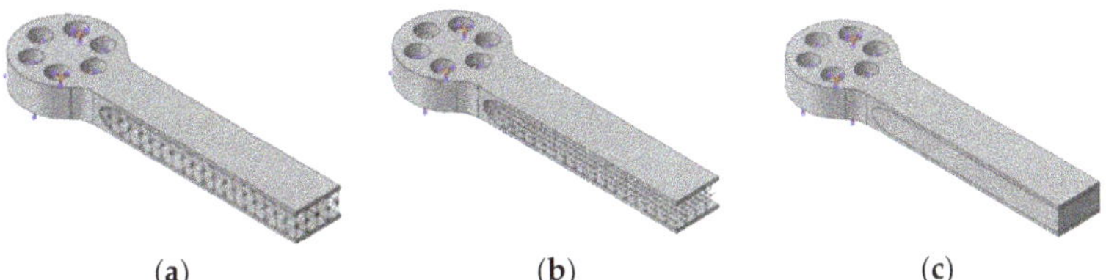

Figure 3. (**a**,**b**) Discrete model of FCC and Diamond lattice structure. (**c**) Homogenized lattice model.

To model the interaction between shaker and specimen, the counterbore and the circular region below the solid are considered to be clamped to ground. A tetrahedral free mesh with 10 nodes quadratic elements (C3D10) is employed to discretize the spatial domain. The concentrated masses were modelled with two reference points kinematically coupled at the tip of the specimen, placed in the center of gravity of the two masses.

The material employed is AlSi10Mg, while the homogenized version of the model includes the elastic properties listed in Table 3.

Table 3. Homogenized mechanical elastic properties and yield strength of the FCC and Diamond unit cells.

Cell Type	ρ [%]	Elastic Modulus [MPa]	Shear Modulus [MPa]	Poisson's Ratio	Yield Stress [MPa]
FCC	7.3	1371.4	314.4	0.19	5.74
Diamond	7.0	174.1	178.8	0.46	1.65

The validation of the numerical models has been performed by comparing the experimental (evaluated through a frequency sweep test) and numerical acceleration response in terms of resonance frequency ω_0 and amplitude $|A(\omega_0)|$. For each model, a structural damping ratio of 0.21% and 0.31% has been considered for the FCC and Diamond. These values represent the mean values of experimental damping ratios measured during the experiments. The FRF of the FE models have been numerically computed imposing a constant acceleration of 1 g in the frequency range between 350 Hz and 450 Hz for the FCC specimen, and 300–400 for the diamond configuration. The results are listed in Table 4, along with the corresponding modeshapes, which are displayed in Figure 4 together with the mean value and standard deviation of the experimental results measured on 10 and 9 specimens for the FCC and Diamond cell respectively. The numerical models slightly overestimate the first natural frequency ω_0 with an error of about 3–4% for the homogenized model increasing to 5–8% for the discrete model. The error in the numerical prediction of the amplitude $|A(\omega_0)|$ is 4–5% without a significant effect of the cell topology or the employed numerical model. These differences can be attributed to the geometrical differences and the variability in the damping that are introduced during the manufacturing process.

Table 4. First natural frequency and FRF of the numerical models and the results of the sweep tests.

		Damping Ratio [%]	f_0 [Hz]	$|A(\omega_0)|$
FCC	Numerical (Homogenized)	0.21	410.7	244.5
	Numerical (Discrete)	0.21	420.8	243.9
	Experimental *	0.21 (0.03)	399.2 (2.7)	256.2 (31.7)
Diamond	Numerical (Homogenized)	0.31	368.7	166.2
	Numerical (Discrete)	0.31	383.8	166.0
	Experimental *	0.31 (0.08)	353.8 (3.8)	176.0 (25.5)

* Mean and standard deviation (in parentheses).

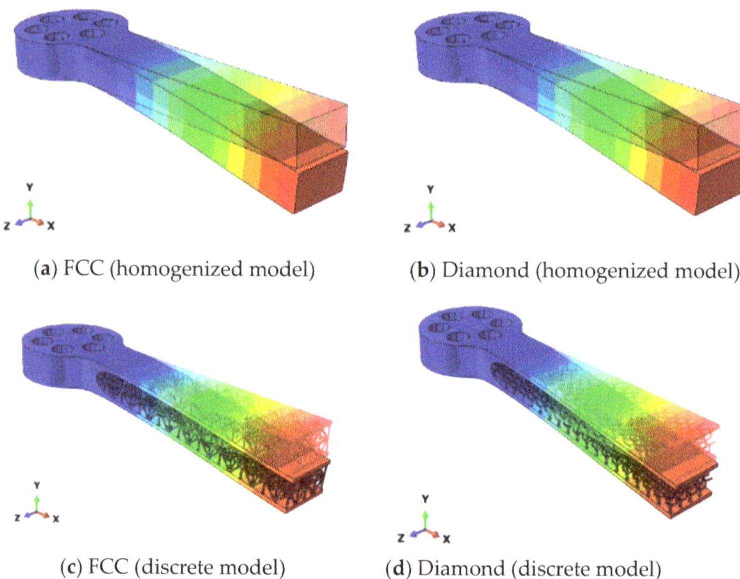

(**a**) FCC (homogenized model)

(**b**) Diamond (homogenized model)

(**c**) FCC (discrete model)

(**d**) Diamond (discrete model)

Figure 4. First shape mode of (**a**,**b**) the homogenized models and (**c**,**d**) the discrete models.

It is worth remarking that the homogenized model is less computationally expensive and the global dynamic behavior in the neighborhood of the first natural frequency is well captured with an acceptable error. On the contrary, the discrete model requires a refined mesh due to the peculiar geometry, but it is fundamental for the analysis of the local failures and stress concentrations.

Numerical Analyses for Failure Identification

A generally accepted method for the failure identification in the context of mechanical systems captures the natural frequency changes in response to a variation of mechanical properties. It is assumed that the failure of crossing struts is responsible for such a change. A graphical example is illustrated in Figure 5 and implemented in the numerical model in terms of broken strut, Figure 5a, and missing node, Figure 5b.

The concept is elucidated in the numerical frequency responses displayed in Figure 5c,d for FCC and Diamond specimens, respectively. As expected, a broken strut (red line) leads to a small frequency change, while a missing node (blue line) affects the resonance frequency in a more evident way, which leads also to a visible reduction also in terms of tip acceleration.

The numerical results of as-designed and induced damage for FCC and Diamond lattice models are reported in Table 5. For the FCC cell, the failure of a strut induces a reduction of about 0.14% and 17.4% of the natural frequency and tip acceleration. A more significant damage is given by a node failure when the frequency and tip acceleration decrease of about 0.5% and 61.9%. For the Diamond cell, a similar trend is visible but with a smaller drop of the frequency and tip acceleration since the strut connectivity is much higher.

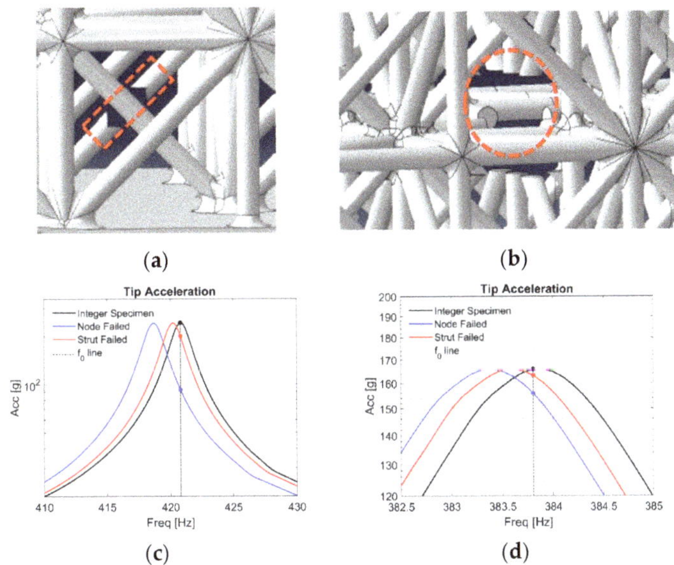

Figure 5. Detailed view of the damage families considered in this work for the FCC lattice configuration. (**a**) Broken strut and (**b**) missing node. (**c**,**d**) FRF of the as-designed (black line) and induced-damage geometry (red and blue lines) for FCC and diamond cell respectively.

Table 5. Reduction of frequency and tip acceleration under resonant conditions through numerical models.

		As-Designed	Missing Strut	Missing Node
FCC	f_0 [Hz]	420.8	420.2	418.7
	Tip acceleration [g]	243.7	201.4	92.9
Diamond	f_0 [Hz]	420.8	420.2	418.7
	Tip acceleration [g]	243.7	201.4	92.9

Based on these numerical results, we set the failure condition corresponding to the failure of a single strut. For the endurance tests under resonant conditions, this condition leads to an acceleration drop of about 15% and 1.5% for the FCC and Diamond specimens, respectively. For the random tests, a frequency reduction of 0.14% and 0.05% is employed.

4. Experimental Results

This section describes the experimental results for the endurance and random tests, which are carried out assuming the criteria of failure identification defined in the previous chapter, i.e., acceleration drops and natural frequency variations for endurance and random test, respectively.

4.1. Endurance Test

A number of lattice specimens have been tested in order to have enough data for the SN diagrams. That is, five FCC prototypes and four for Diamond specimens. The results for the FCC and Diamond samples are reported in Tables 6 and 7, in terms of tip displacement, excitation frequency, failure time, and number of cycles to failure (dependent from failure time and excitation frequency).

Table 6. Endurance test results for the FCC specimen.

Specimen No.	Tip Displacement [mm]	Excitation Frequency [Hz]	Failure Time [s]	Cycle to Failure
7	0.31	398	88.2	35104
8	0.3	400	99.2	39680
9	0.32	392.5	71.4	28024
10	0.2	400.4	598.1	239480
11	0.18	396.3	610.9	242100

Table 7. Endurance tests results for the Diamond specimen.

Specimen No.	Tip Displacement [mm]	Excitation Frequency [Hz]	Failure Time [s]	Cycle to Failure
1	0.23	352.6	236.1	83249
2	0.25	357.3	421.1	150460
3	0.33	353.9	55.4	19606
4	0.32	349.6	102.2	35729

As mentioned in the previous section, acceleration drops of about 15% and 1.5% are considered sufficiently high to be representative to a failure condition. A representative example is shown in Figure 6 in terms of time history for FCC and Diamond specimens (number 7 and 4).

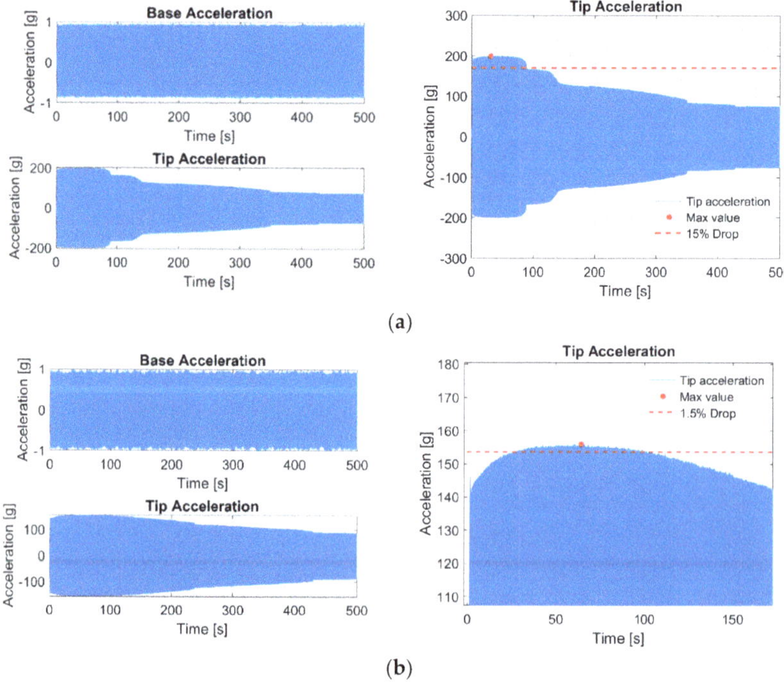

Figure 6. Accelerations measured in correspondence of the base and tip during the endurance tests. (**a**) FCC specimen number 7 and (**b**) Diamond specimen 4.

As expected, the tip acceleration for FCC specimens, displayed in Figure 6a, suddenly decreases after 88 s approximately, which is a blueprint of natural frequency decrease. According to the above arguments, a 15% reduction of the tip acceleration is representative of a variation in natural frequency caused by the failure of the most loaded strut within the lattice. A similar dynamic response is observed also for the FCC specimen number 8 and 9. In contrast, a different behavior is noticed for the Diamond, as shown in Figure 6b. That is, the Diamond specimens exhibit a smoother acceleration drop. For the sample number 4, the failure time is 102.2 s at 349.6 Hz (35,729 cycles), corresponding to a tip displacement of about 0.32 mm.

As expected, Tables 6 and 7 show that the fatigue life increases as the tip displacement decreases, which corresponds to lower stresses within the lattice and, therefore, a longer fatigue life. The results discussed within this section will be used later in the paper to validate the SN diagram representative of the lattice metamaterial, which is necessary to also estimate the fatigue life under random excitation.

4.2. Random Tests

In contrast with the endurance tests, here, the failure identification is based on a frequency drop of about 0.14% and 0.05% the nominal value for the FCC and diamond prototypes, above which the lattice sandwich is considered as damaged. Figure 7a,b and Figure 8a,b are the spectrograms that show the frequency evolution in time relative to the FCC specimen number 4 and Diamond specimen number 7. Here, the frequency response function estimator H1 is computed windowing the time interval with a moving time window and displayed in the figures with a colored surface. The natural frequency evolution is tracked and reported with a red line in Figures 7c and 8c. The failure time is identified as the intersection between the red line and the horizontal dashed blue line, which corresponds to the threshold value representative of the failure computed numerically.

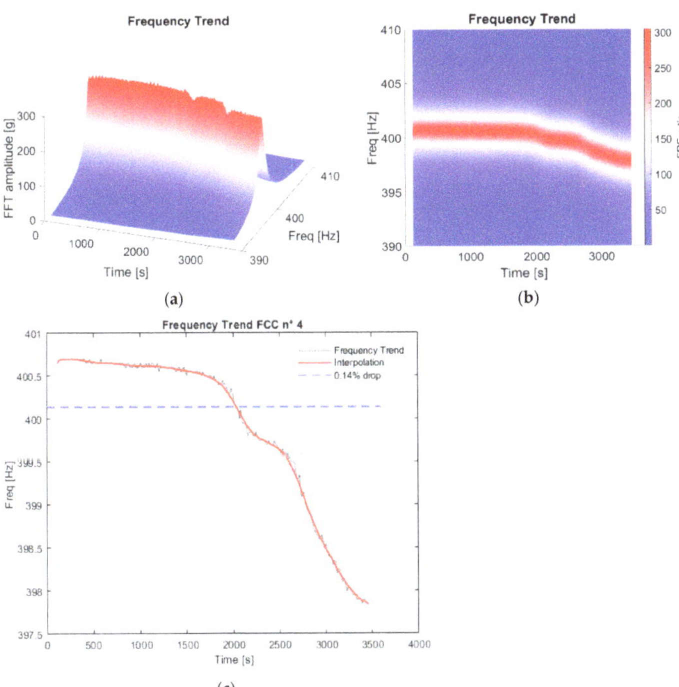

Figure 7. Spectrogram relative to the FCC specimen number 4 tested under random vibration. (**a**) Axonometric and (**b**) top view. The frequency trend is tracked and displayed in (**c**) for a better visualization of the failure time.

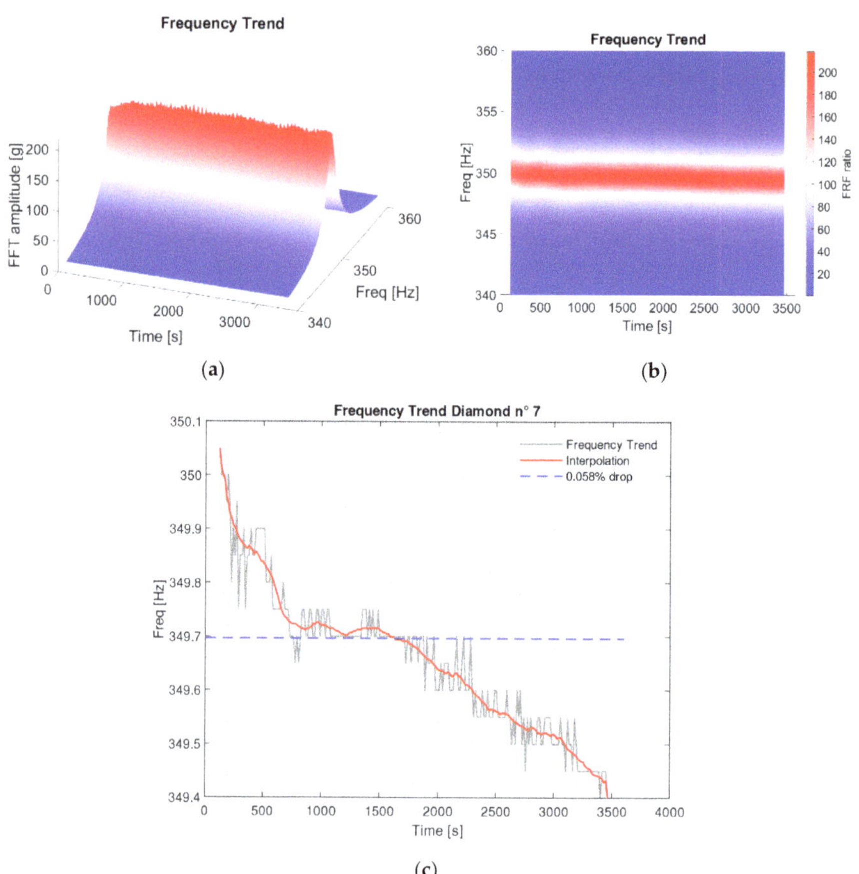

Figure 8. Spectrogram relative to the Diamond specimen number 7 tested under random vibration. (**a**) Axonometric view and (**b**) top view. (**c**) Peak amplitude tracked in time to show the frequency drop.

The FCC specimen number 4 fails after 2035 s, after which there is a further natural frequency decrease until the end of the test. The specimen number 3 and 4 have been tested through a PSD with reduced amplitude (10 times lower), while specimen number 2 and 5 have been tested with 1/15 of the nominal amplitude, see Table 2. The experimental results for the number of tested specimens are reported in Tables 8 and 9, in terms of failure time and frequency drops.

Table 8. Random test results for the FCC lattice specimens.

Specimen No.	PSD Type	Test Duration [h]	Initial Sweep Freq. [Hz]	Mid-Sweep Freq. [Hz]	Final Sweep Freq. [Hz]	Initial FRF	Time to Failure [s]
2	/15	2	398.9	398.1	397.5	273.5	2400
3	/10	1	399.9	/	398.5	198.2	2060
4	/10	1	400.8	/	397	259.2	2035
5	/15	2	401.4	401.4	400.5	273.8	4355

Table 9. Random tests result for the Diamond lattice specimens.

Specimen No.	PSD Type	Test Duration [h]	Initial Sweep Freq. [Hz]	Mid-Sweep Freq. [Hz]	Final Sweep Freq. [Hz]	Initial FRF	Time to Failure [s]
6	/10	1	358.9	/	357.8	122.6	1620
7	/10	1	351.9	/	350.2	169.3	855
8	/15	2	353.7	352.9	352.6	172.8	Undamaged
9	/15	2	357.1	356.3	355.9	151.8	2430

5. Fatigue Life Analysis

Now, the fatigue behavior of lattice structures tested under resonant conditions and random dynamic loads is studied by combining the experimentally computed number of cycles to failure, along with the numerical stress computed through the FE models.

The discrete models, see Figure 3, are employed to estimate the stresses within the lattice sandwich. That is, a number of FE analyses are performed to map the stresses as a function of the tip displacement (relative to the base) and damping ratio. The state of stress is computed through the numerical procedure presented in Ref. [26], which allows to evaluate an equivalent stress amplitude based on the application of the Sines criterion for each surface node i:

$$\sigma_{a,eq}^i = \frac{\sigma_{VM,a}^i}{1 - \frac{I_{1,m}^i}{Rm}}$$

where $\sigma_{VM,a}^i$ and $I_{1,m}^i$ are the equivalent Von Mises stress amplitude and the first invariant. The map of the maximum equivalent stress amplitude in the lattice structure, displayed in Figure 9, has been obtained as interpolation of numerical results obtained for different values of imposed base acceleration and damping ratios.

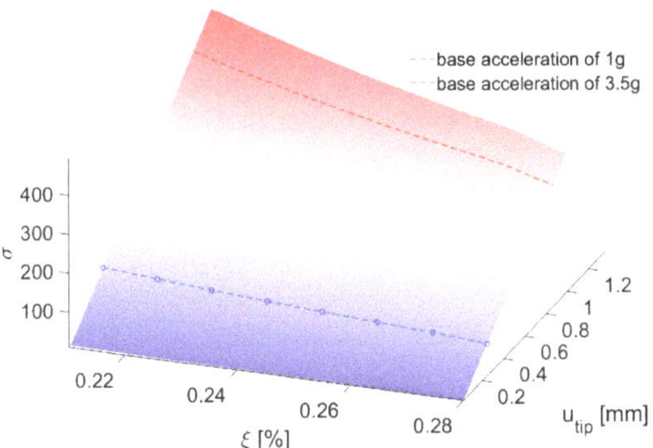

Figure 9. Equivalent stress amplitude map for FCC specimens.

5.1. Fatigue Life Evaluation under Resonant Conditions

The tip displacement and the damping ratio measured during the experimental tests under resonant conditions are introduced as input in Figure 9, which returns the associated stress amplitude level. Figure 10 shows the position of such points in the SN diagram of the bulk material in as-built condition, obtained from previous tests performed on solid specimens printed in the vertical direction with a loading ratio R = −1 [33].

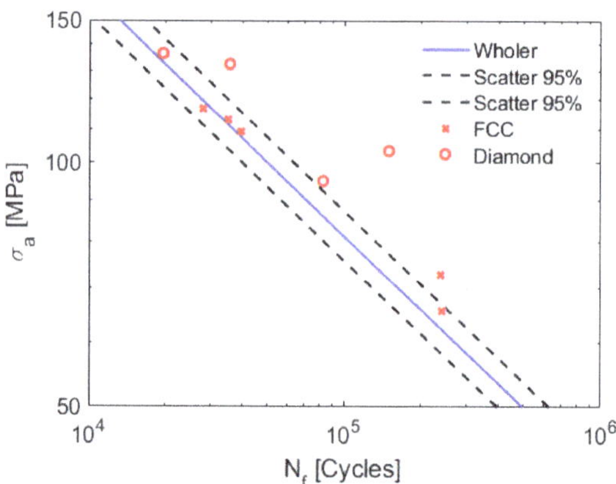

Figure 10. SN curve of the AlSi10Mg bulk material with superimposed SN data points associated to the FCC and Diamond configuration for resonant condition tests.

Regarding the FCC configuration, four points lie within the 95% scatter band, while only one point is located outside. In contrast, for the Diamond configuration the data points are more scattered and, some of them, fall outside the 95% scatter band. This behavior can be attributed to several factors. Firstly, the time histories of the endurance tests are characterized by a less regular trend: at the beginning of the test, the acceleration increases and later decreases, which can be due to nonlinear interactions, see Figure 6. This modifies the actual starting point (with maximum amplitude) of the analysis. The differences can be also attributed to the manufacturing process, which introduces high variability in the printed dimension, especially in correspondence of the crossing points between lattice struts. In this context, a mismodelling of the fillet radius can be a source of errors in the stress estimation, and therefore, in the SN data points.

Despite these discrepancies, the SN curve of bulk material, combined with the numerical model can be considered an effective method to predict the fatigue life of lattice structures through standard fatigue design criteria.

5.2. Damage Evaluation under Random Fatigue Loads

To compute the equivalent stress, a Rainflow-counting algorithm is applied to the measured acceleration and, therefore, the corresponding displacement [34]. The algorithm computes a matrix with the alternate displacement as well as the relative occurrence. Such information is employed to evaluate the equivalent stress amplitude (through the map in Figure 9) which, when combined with the number of occurrences, are the necessary data to compute the cumulative damage to the FCC and Diamond lattice specimens using Miner's rule and the SN diagram reported in Figure 10. According to [33], the fatigue limit of bulk material is set equal to 50 MPa. Below this value, the Haibach assumption is used for damage evaluation. Results are reported in Tables 10 and 11.

Table 10. Cumulative damage computed by Miner's rule for the FCC specimens.

Specimen No.	2	3	4	5
Time to failure [s]	2400	2060	2035	4355
Damage to failure	0.9	0.94	1.55	1.31

Table 11. Cumulative damage computed by Miner's rule for the Diamond specimens.

Specimen No.	6	7	8	9
Time to failure [s]	1620	855	/	2430
Damage to failure	1.37	0.93	/	1.18

Some considerations follow. The use of dynamic models with the Miner's rule predicts cumulative damage between 0.9 and 1.55 for the FCC specimens, while for the Diamond lattice between 0.93 and 1.37. The theoretical failure occurs when the damage reaches 1, considering the range reported in the table, the evaluation of fatigue damage can be considered acceptable. These results are also in line with values obtained in [35].

6. Conclusions

In this study, the fatigue behavior of lattice structures under random cyclic loads has been numerically and experimentally investigated, providing a guideline for the assessment of lattice structures. The analysis is performed onto two distinct cell topologies, i.e., FCC and Diamond-based, which have been experimentally studied through different testing procedures, i.e., sweep, endurance and random tests. The modification in the dynamic response of the lattice sandwich during the experimental tests have been simulated through finite element analyses of the structure with induced defects (broken strut or missing node). Based on these numerical results, failure in vibration fatigue tests is defined as the acceleration change or frequency drop as induced by the failure of a single strut.

Based on the SN diagram of the bulk material, the fatigue life of FCC and Diamond specimens under resonant conditions is predicted with good accuracy by defining an equivalent stress amplitude as a function of the tip displacement and damping. Under random vibrations, fatigue damage is accomplished by using the SN curve of the bulk material, the Rainflow's algorithm and Miner's rule. The predicted results have been shown to be consistent with the failure observed during the tests, with damage values in the range between 0.9–1.55 and 0.93–1.37 for the FCC and Diamond lattice specimens.

Author Contributions: M.P.: formal analysis, investigation, writing—original draft, visualization; M.G.C.: formal analysis, investigation; S.B.: conceptualization, writing—review and editing, funding acquisition; E.R.: investigation, writing—original draft; F.B.: conceptualization, resources; S.F.: conceptualization, writing—review and editing, supervision, project administration. All authors have read and agreed to the published version of the manuscript.

Funding: This research received no external funding.

Institutional Review Board Statement: Not applicable.

Informed Consent Statement: Not applicable.

Data Availability Statement: The data presented in this study are available on request from the corresponding author.

Acknowledgments: The Italian Ministry of Education, University and Research is acknowledged for the support provided through the "Department of Excellence LIS4.0—Lightweight and Smart Structures for Industry 4.0" Project. Material for experiments was supplied by Leonardo Electronics S.p.A. Their support of research activities is greatly acknowledged.

Conflicts of Interest: The authors declare no conflict of interest.

References

1. Tancogne-Dejean, T.; Diamantopoulou, M.; Gorji, M.B.; Bonatti, C.; Mohr, D. 3D Plate-Lattices: An Emerging Class of Low-Density Metamaterial Exhibiting Optimal Isotropic Stiffness. *Adv. Mater.* **2018**, *30*, 1–6. [CrossRef]
2. Kolken, H.M.A.; Zadpoor, A.A. Auxetic mechanical metamaterials. *RSC Adv.* **2017**, *7*, 5111–5129. [CrossRef]
3. Scalzo, F.; Totis, G.; Vaglio, E.; Sortino, M. Experimental study on the high-damping properties of metallic lattice structures obtained from SLM. *Precis. Eng.* **2021**, *71*, 63–77. [CrossRef]

4. Tancogne-Dejean, T.; Spierings, A.B.; Mohr, D. Additively-manufactured metallic micro-lattice materials for high specific energy absorption under static and dynamic loading. *Acta Mater.* **2016**, *116*, 14–28. [CrossRef]
5. Maconachie, T.; Leary, M.; Lozanovski, B.; Zhang, X.; Qian, M.; Faruque, O.; Brandt, M. SLM lattice structures: Properties, performance, applications and challenges. *Mater. Des.* **2019**, *183*, 108137. [CrossRef]
6. Lira, C.; Scarpa, F.; Rajasekaran, R. A gradient cellular core for aeroengine fan blades based on auxetic configurations. *J. Intell. Mater. Syst. Struct.* **2011**, *22*, 907–917. [CrossRef]
7. Spadoni, A.; Ruzzene, M.; Scarpa, F. Dynamic response of chiral truss-core assemblies. *J. Intell. Mater. Syst. Struct.* **2006**, *17*, 941–952. [CrossRef]
8. Colombo, C.; Biffi, C.; Fiocchi, J.; Scaccabarozzi, D.; Saggin, B.; Tuissi, A.; Vergani, L. Modulating the damping capacity of SLMed AlSi10Mg trough stress-relieving thermal treatments. *Theor. Appl. Fract. Mech.* **2020**, *107*, 1–6. [CrossRef]
9. Fiocchi, J.; Biffi, C.A.; Scaccabarozzi, D.; Saggin, B.; Tuissi, A. Enhancement of the Damping Behavior of Ti6Al4V Alloy through the Use of Trabecular Structure Produced by Selective Laser Melting. *Adv. Eng. Mater.* **2020**, *22*, 1–6. [CrossRef]
10. Liu, J.; Guo, K.; Sun, J.; Sun, Q.; Wang, L.; Li, H. Compressive behavior and vibration-damping properties of porous Ti-6Al-4V alloy manufactured by laser powder bed fusion. *J. Manuf. Process.* **2021**, *66*, 1–10. [CrossRef]
11. Zadeh, M.N.; Alijani, F.; Chen, X.; Dayyani, I.; Yasaee, M.; Mirzaali, M.J.; Zadpoor, A.A. Dynamic characterization of 3D printed mechanical metamaterials with tunable elastic properties. *Appl. Phys. Lett.* **2021**, *118*, 211901. [CrossRef]
12. Benedetti, M.; du Plessis, A.; Ritchie, R.; Dallago, M.; Razavi, S.; Berto, F. Architected cellular materials: A review on their mechanical properties towards fatigue-tolerant design and fabrication. *Mater. Sci. Eng. R Rep.* **2021**, *144*, 100606. [CrossRef]
13. Yavari, S.A.; Ahmadi, S.; Wauthle, R.; Pouran, B.; Schrooten, J.; Weinans, H.; Zadpoor, A.A. Relationship between unit cell type and porosity and the fatigue behavior of selective laser melted meta-biomaterials. *J. Mech. Behav. Biomed. Mater.* **2015**, *43*, 91–100. [CrossRef]
14. Van Hooreweder, B.; Apers, Y.; Lietaert, K.; Kruth, J.-P. Improving the fatigue performance of porous metallic biomaterials produced by Selective Laser Melting. *Acta Biomater.* **2017**, *47*, 193–202. [CrossRef] [PubMed]
15. Ahmadi, S.; Hedayati, R.; Li, Y.; Lietaert, K.; Tümer, N.; Fatemi, A.; Rans, C.; Pouran, B.; Weinans, H.; Zadpoor, A.; et al. Fatigue performance of additively manufactured meta-biomaterials: The effects of topology and material type. *Acta Biomater.* **2018**, *65*, 292–304. [CrossRef]
16. Zhao, S.; Li, S.; Wang, S.; Hou, W.; Li, Y.; Zhang, L.; Hao, Y.; Yang, R.; Misra, R.; Murr, L. Compressive and fatigue behavior of functionally graded Ti-6Al-4V meshes fabricated by electron beam melting. *Acta Mater.* **2018**, *150*, 1–15. [CrossRef]
17. Yavari, S.A.; Wauthle, R.; van der Stok, J.; Riemslag, A.; Janssen, M.; Mulier, M.; Kruth, J.; Schrooten, J.; Weinans, H.; Zadpoor, A.A. Fatigue behavior of porous biomaterials manufactured using selective laser melting. *Mater. Sci. Eng. C* **2013**, *33*, 4849–4858. [CrossRef] [PubMed]
18. Abad, E.M.K.; Khanoki, S.A.; Pasini, D. Fatigue design of lattice materials via computational mechanics: Application to lattices with smooth transitions in cell geometry. *Int. J. Fatigue* **2013**, *47*, 126–136. [CrossRef]
19. Dobson, S.D.; Starr, T.L. Powder characterization and part density for powder bed fusion of 17-4 PH stainless steel. *Rapid Prototyp. J.* **2020**, *27*, 53–58. [CrossRef]
20. Donik, Č.; Kraner, J.; Paulin, I.; Godec, M. Influence of the Energy Density for Selective Laser Melting on the Microstructure and Mechanical Properties of Stainless Steel. *Metals* **2020**, *10*, 919. [CrossRef]
21. Ullah, A.; Wu, H.; Rehman, A.U.; Zhu, Y.; Liu, T.; Zhang, K. Influence of laser parameters and Ti content on the surface morphology of L-PBF fabricated Titania. *Rapid Prototyp. J.* **2020**, *27*, 71–80. [CrossRef]
22. Pal, S.; Lojen, G.; Gubeljak, N.; Kokol, V.; Drstvensek, I. Melting, fusion and solidification behaviors of Ti-6Al-4V alloy in selective laser melting at different scanning speeds. *Rapid Prototyp. J.* **2020**, *26*, 1209–1215. [CrossRef]
23. Pawlak, A.; Szymczyk-Ziółkowska, P.; Kurzynowski, T.; Chlebus, E. Selective laser melting of magnesium AZ31B alloy powder. *Rapid Prototyp. J.* **2019**, *26*, 249–258. [CrossRef]
24. Liang, C.; Hu, Y.; Liu, N.; Zou, X.; Wang, H.; Zhang, X.; Fu, Y.; Hu, J. Laser Polishing of Ti6Al4V Fabricated by Selective Laser Melting. *Metals* **2020**, *10*, 191. [CrossRef]
25. Dallago, M.; Raghavendra, S.; Luchin, V.; Zappini, G.; Pasini, D.; Benedetti, M. The role of node fillet, unit-cell size and strut orientation on the fatigue strength of Ti-6Al-4V lattice materials additively manufactured via laser powder bed fusion. *Int. J. Fatigue* **2020**, *142*, 105946. [CrossRef]
26. Boniotti, L.; Beretta, S.; Patriarca, L.; Rigoni, L.; Foletti, S. Experimental and numerical investigation on compressive fatigue strength of lattice structures of AlSi7Mg manufactured by SLM. *Int. J. Fatigue* **2019**, *128*, 105181. [CrossRef]
27. Yang, L.; Yan, C.; Cao, W.; Liu, Z.; Song, B.; Wen, S.; Zhang, C.; Shi, Y.; Yang, S. Compression–compression fatigue behaviour of gyroid-type triply periodic minimal surface porous structures fabricated by selective laser melting. *Acta Mater.* **2019**, *181*, 49–66. [CrossRef]
28. Kelly, C.N.; Francovich, J.; Julmi, S.; Safranski, D.; Guldberg, R.E.; Maier, H.J.; Gall, K. Fatigue behavior of As-built selective laser melted titanium scaffolds with sheet-based gyroid microarchitecture for bone tissue engineering. *Acta Biomater.* **2019**, *94*, 610–626. [CrossRef]
29. Bobbert, F.; Lietaert, K.; Eftekhari, A.A.; Pouran, B.; Ahmadi, S.; Weinans, H.; Zadpoor, A.A. Additively manufactured metallic porous biomaterials based on minimal surfaces: A unique combination of topological, mechanical, and mass transport properties. *Acta Biomater.* **2017**, *53*, 572–584. [CrossRef] [PubMed]

30. Richard, B.; Pellicone, D.; Anderson, W.G. Progress on the Development of a 3D Printed Loop Heat Pipe. In Proceedings of the 2019 35th Semiconductor Thermal Measurement, Modeling and Management Symposium (SEMI-THERM), San Jose, CA, USA, 18–22 March 2019; pp. 1–11.
31. Schmitz, A.; Horst, P. A finite element unit-cell method for homogenised mechanical properties of heterogeneous plates. *Compos. Part A Appl. Sci. Manuf.* **2014**, *61*, 23–32. [CrossRef]
32. Tancogne-Dejean, T.; Mohr, D. Elastically-isotropic truss lattice materials of reduced plastic anisotropy. *Int. J. Solids Struct.* **2018**, *138*, 24–39. [CrossRef]
33. Sausto, F.; Carrion, P.; Shamsaei, N.; Beretta, S. Fatigue failure mechanisms for AlSi10Mg manufactured by L-PBF under axial and torsional loads: The role of defects and residual stresses. *Int. J. Fatigue* **2021**, submit.
34. Lee, Y.L.; Tjhung, T. *Rainflow Cycle Counting Techniques*; Elsevier Inc.: Amsterdam, The Netherlands, 2012. [CrossRef]
35. Nourian-Avval, A.; Fatemi, A. Variable amplitude fatigue behavior and modeling of cast aluminum. *Fatigue Fract. Eng. Mater. Struct.* **2021**, *44*, 1611–1621. [CrossRef]

Article

The Use of Surface Topography for the Identification of Discontinuous Displacements Due to Cracks

Fatih Uzun and Alexander M. Korsunsky * 

Multi-Beam Laboratory for Engineering Microscopy (MBLEM), Department of Engineering Science, University of Oxford, Oxford OX1 3PJ, UK; fatih.uzun@eng.ox.ac.uk
* Correspondence: alexander.korsunsky@eng.ox.ac.uk; Tel.: +44-1865-2-73043

Received: 9 July 2020; Accepted: 28 July 2020; Published: 2 August 2020

Abstract: The determination of three components of displacements at material surfaces is possible using surface topography information of undeformed (reference) and deformed states. The height digital image correlation (hDIC) technique was developed and demonstrated to achieve micro-level in-plane resolution and nanoscale out-of-plane precision. However, in the original formulation hDIC and other topography-based correlation techniques perform well in the determination of continuous displacements. In the present study of material deformation up to cracking and filan failure, the ability to identify discontinuous triaxial displacements at emerging discontinuities is important. For this purpose, a new method reported herein was developed based on the hDIC technique. The hDIC solution procedure comprises two stages, namely, integer-pixel level correlation and sub-pixel level correlation. In order to predict the displacement and height changes in discontinuous regions, a smoothing stage was inserted between the two main stages. The proposed method determines accurately the discontinuous edges, and the out-of-plane displacements become sharply resolved without any further intervention in the algorithm function. High computational demand required to determine discontinuous displacements using high density topography data was tackled by employing the graphics processing unit (GPU) parallel computing capability with the paging approach. The hDIC technique with GPU parallel computing implementation was applied for the identification of discontinuous edges in an aluminium alloy dog bone test specimen subjected to tensile testing up to failure.

Keywords: surface topography; optical profilometry; height digital image correlation; discontinuous displacements; triaxial displacements

1. Introduction

The use of the digital image correlation (DIC) technique for the determination of biaxial displacements dates back to the 1980s. After the introduction of this technique by Parks and Vincent [1] for measuring displacements using speckle photography, Chu et al. [2] applied digital speckle patterns in the context of experimental mechanics of solids. Today, this technique is widely used for monitoring biaxial displacements at the surface of samples with 2D or 3D geometries as presented in the studies of Hild and Roux [3] and Zhao et al. [4]. The missing information related to the third component of displacements (out-of-plane with respect to the sample surface tangent plane) was typically neglected, because digital imaging cameras were not capable of recording information about the out-of-plane displacements at the surface with micro-level in-plane resolution of the samples being investigated. Current topography based DIC techniques determine out-of-plane displacements with very high-resolution over small areas [5–10]. On the other hand, 3D DIC techniques provide depth information with very low resolution and low precision over large areas [11–14]. The height digital image correlation (hDIC) [15] which is a true full field method and based on focus stacking optical microscopy (FSOM), fills the gap between the two approaches. Beeck et al. [16] also used confocal microscopy for obtaining topography information to perform DIC, but this scanning method is intrinsically slow [8] and requires complex finite element calculations and, accordingly, it is far from being a full field method. However, all topography based DIC techniques, including the hDIC, have been optimised for determination of continuous displacements.

The displacement fields arising in engineering components can be investigated in situ using DIC. The low cost and high accuracy of the technique eliminate the need for expensive laboratory strain measurement equipment, such as strain gauges or extensometers. Sause [17] provided improvements in the resolution of digital image recording techniques that allowed capturing small defects and instances of failure, such as cracks not visible to the naked eye. Abanto-Bueno and Lambros [18] used DIC for the purpose of investigating crack growth, Mathieu [19] proposed a crack propagation law, Mokhtarishirazabad et al. [20] evaluated crack-tip fields, and Hamam et al. [21] and Mcneill et al. [22] estimated stress intensity factors, e.g., in the investigation of mode I crack propagation as it was presented by Tahreer et al. [23]. These studies performed correlation for continuous displacement fields but did not provide information regarding the discontinuous edges associated with the cracks. Chernyatin et al. [24] developed a mathematical and numerical correction method for displacements determined by DIC. The proposed model provides a theoretical description of the discontinuous displacement fields associated with material edges. However, this estimation process begins with the elimination of a rectangular area. Therefore, the interpretation is not based on real experimental measurements. Jandejsek et al. [25] performed standard fracture toughness tests to determine in-plane stress and strain distributions around a crack using DIC. However, the solutions include smoothing stages and do not capture the detail of discontinuous displacements at material edges within the crack. Hosdez et al. [26] studied crack growth in ductile cast iron using DIC and the potential drop method. The authors visually presented the displacement distribution associated with the crack at different resolutions, with the increasing resolution providing more details of the precise crack shape and deformation fields. The length of the crack was determined with good precision; however, the information about the displacements and strains around the crack was missing from the report. Bourdin et al. [27] presented the Heaviside-DIC technique for the measurement of plastic localisation in polycrystalline metallic materials. The proposed method focuses on the determination of discontinuous deformations formed by slip bands but requires additional treatment of the subsets. Cinar et al. [28] also used Heaviside-DIC technique, but correlation around the crack fails within a wide range and results in errors that need to be rectified by extrapolation. As a consequence, current DIC techniques for the determination of discontinuous deformations use pixel intensities with algorithms specified to the applied problem. These methods were not developed based on surface topography information and accordingly they are not able to determine out-of-plane displacements in and around deformed sections. Final representations of discontinuous edges are blurry images that need guiding lines.

In spite of the fact that conventional DIC techniques are able to determine the discontinuous displacements associated with material edges such as cracks, they are not able to determine the boundaries of discontinuous displacements. Thus, representations of discontinuous displacements do not include real boundaries, but rather visualise them as blurry (smeared out, or smoothed) images of fracture zones. In addition to the lack of details of discontinuous edges, the depth (out-of-plane displacements) components cannot be determined using conventional DIC techniques. The lack of information about the depth component of displacements prevents the calculation of the Mode III component of crack-related deformation fields.

The hDIC technique [15] was proposed recently for the correlation of optical profilometry data obtained using "infinite focus" microscopy. This allows the use of height data instead of the conventional grey scale colour intensity in typical digital images for the purpose of determination of triaxial surface displacements. The hDIC technique was developed as a two-stage process, comprising integer-pixel level and sub-pixel level correlation stages. The integer-pixel level stage determines the best matching pixels to determine large shifts, while the sub-pixel level correlation stage searches for the best matching displacement values using interpolation. In this two-stage process, the role of integer-pixel level cross-correlation is to determine the best matching pixel as a starting point for the minimisation process conducted during the sub-pixel level correlation. The determination of large scale displacements using this two-stage process was previously validated by investigating displacements during the tensile test by the present authors [15] and by comparing strains calculated using the hDIC displacement results with synchrotron diffraction measurements by Uzun et al. [29].

After the completion of integer-pixel level cross-correlation and sub-pixel level correlation stages of the hDIC technique, the results are presented and analysed following the smoothing that plays an important role in the digital image correlation algorithms serving the purpose of eliminating noise, as it was explained in the studies of Craven and Wahba [30] and Woltring [31]. As a consequence, only continuous displacement fields can be extracted as it was stated in the studies of Palanca [32] and Woltring [31]. Li et al. [33] stated that careful adjustment of the smoothing parameters is crucial, because the strain calculation is highly sensitive to displacement noise. In the case of discontinuous displacement fields, smoothing has a harmful effect on the determination of discontinuous edges. In order to prevent the removal of discontinuous edges by smoothing, several methods were developed in [25], and by Mathieu et al. [19] and Réthoré et al. [34] to guide the DIC algorithm in describing the displacement fields around cracks (but not across them). However, in a generic analytical approach, the determination of discontinuous displacements should be performed at any surface without additional guiding procedures. The current methods do not satisfy this demand.

In this study, topography data collected using the optical profilometry technique were used for the first time to determine defects and failures that create discontinuous displacements and height profile changes. For this purpose, the hDIC technique was modified as an automatically working algorithm that does not need additional guiding property around the discontinuous edges. A new three-stage correlation process is presented to determine the displacements accurately at discontinuous edges. After integer-pixel level cross-correlation stage, a smoothing function is fit with displacement results of the first stage to predict displacements in damaged spaces of the target topography data. The matching points predicted in the target topography are used as the centre point of the sub-pixel level correlation stage. The use of topography data allowed the determination of the height profile of damaged sections in the experiments conducted to correlate initial and after break states of a tensile test specimen.

The use of high-density topography data to achieve high resolution at discontinuous edges increases the computation power demand drastically. In order to deal with this problem, GPU implementation of the hDIC technique was developed based on the paging of subsets corresponding to the reference and target conditions. The gain in speed due to the GPU implementation is presented with benchmark tests.

2. The hDIC Technique with GPU Implementation

The hDIC technique was developed as a two-stage process for the determination of continuous displacement fields. Integer-pixel level cross-correlation and sub-pixel level correlation stages are followed by smoothing to eliminate displacement noise and strain calculations. In order to determine discontinuous displacements without modifying the main form of this correlation process, the smoothing stage was shifted to the middle of two main stages of the hDIC technique.

Resolution of the representation of discontinuous edges is improved by the use of high-density topography data, but this increases the solution time. For the purpose of dealing with this problem, GPU implementation of the hDIC technique was developed by defining subsets of reference and target conditions as pages of three-dimensional arrays as illustrated in Figure 1 for two stages of the hDIC technique.

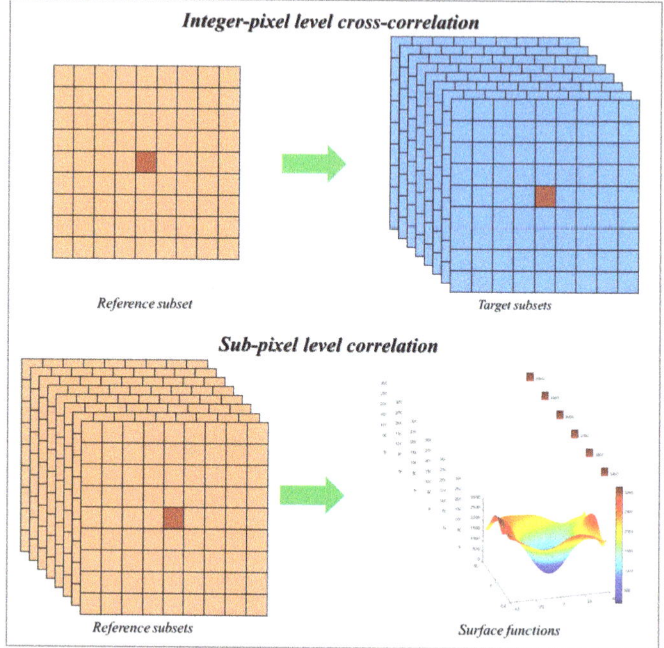

Figure 1. GPU parallel computing implementation of two stages of the height digital image correlation (hDIC) technique.

Integer-pixel level cross-correlation stage aims to determine the best matching pixels between the topography data of reference and target conditions. Dimensions of subsets are formulated as $(2N+1) \times (2N+1)$ where N is an odd number in order to keep a single pixel in the centre of the subset. Cross-correlation is performed using the zero-mean normalised cross-correlation method which is formulated in Equation (1). In this equation, C_i denotes the cross-correlation coefficient of ith subset, $R(x, y)$ denotes the intensity of the pixel at coordinates (x, y) in the reference subset, $T(x', y')$ denotes the intensity of the pixel at coordinates (x', y') in the target subset, R_m denotes the mean intensity of the pixels in the reference subset and T_m is the mean intensity of the pixels in the target subset. Three-dimensional array of target subsets is stored in the GPU memory to perform a parallel calculation of the zero-mean normalised cross-correlation coefficient of each reference subset with all target subsets.

$$C_i = \frac{\sum (R(x, y) - R_m)(T(x', y') - T_m)}{\sqrt{\sum (R(x, y) - R_m)^2} \sqrt{\sum (T(x', y') - T_m)^2}} \tag{1}$$

Subsequent to the determination of the matching subsets, displacements are determined using the final coordinates of the matching pixels using Equation (2) where u denotes the displacement in x-axis and v denotes the displacement in y-axis.

$$x' = x + u(x, y) \tag{2a}$$

$$y' = y + v(x, y) \tag{2b}$$

After the integer-pixel level coarse-fine matching process, displacement noise for x- and y-components (in-plane) of displacements are removed by the bi-cubic B-form least-squares spline interpolation process. The interpolation scheme is employed using the spline function in terms of the weighed sum of B-splines as it is given in Equation (3) where $B_{i,k}(x)$ is the ith and $B_{j,k}(y)$ is the jth B-splines with a degree of $(k-1)$ in x- and y-axes, n_x and n_x are the number of control points in x- and y-axes and $\alpha_{i,j}$ and $\beta_{i,j}$ are the coefficients in the dimensions of i and j. Smoothness of the spline function is determined by adjusting the number of control points. After the smoothing process, matching pixels in the target condition are updated with the coordinates calculated using the smoothed displacements.

$$u(x,y) = \sum_{i=1}^{n_{xy}} \sum_{j=1}^{n_{yx}} \alpha_{i,j} B_{i,k}(x) B_{j,k}(y) \tag{3a}$$

$$v(x,y) = \sum_{i=1}^{n_{xy}} \sum_{j=1}^{n_{yx}} \beta_{i,j} B_{i,k}(x) B_{j,k}(y) \tag{3b}$$

In order to reduce the computation power demand, sub-pixel level correlation is accomplished around the matching pixels in the target condition. The search areas in the target condition are defined as separate cost functions using bi-cubic interpolation method which is given in Equation (4) where $T_s(x',y')$ is the intensity at coordinates (x',y') in the target and $\omega_{i,j}$ is the coefficient at the dimensions of i and j. The final equation, which has continuous derivatives [35], is distributed on a grid surface of unit squares.

$$T_s(x',y') = \sum_{i=0}^{3} \sum_{j=0}^{3} \omega_{i,j} (x')^i (y')^j \tag{4}$$

Functions of the subsets of the target condition are reformulated as cost functions for sub-pixel level correlation process. The cost function of each target subset is determined using the interpolation function and the corresponding reference pixel, separately, using Equation (5) provided below.

$$J(x',y') = (T_s(x',y') - R(x,y))^2 \tag{5}$$

The gradient descent minimisation process simultaneously updates the x' and y' coordinates in the surface, determined by the bi-cubic interpolation function, along the steepest descent direction using Equation (6) until the minimum is achieved. In this equation, γ represents the step size which is kept constant throughout the process. The GPU implementation of this process performs the creation of the cost functions and the gradient descent minimisation process for each reference subset, which are stored as a three-dimensional array in the GPU memory, in parallel.

$$x' = x' - \gamma \frac{\partial}{\partial x'} J(x',y') \tag{6a}$$

$$y' = y' - \gamma \frac{\partial}{\partial y'} J(x',y') \tag{6b}$$

3. Identification of Discontinuous Displacements

The hDIC technique was previously used to measure displacements on the aluminium tensile test specimen by the present authors [15]. The displacement measurements corresponding to the elastic region of the tensile test allowed calculation of Young's modulus and Poisson's ratio. Calculated elastic material properties and the linearity of the longitudinal component of the displacements validated the measurements of the hDIC technique. Smoothed displacement measurements in the plastic region were used to calculate strain distribution. However, that procedure of the hDIC did not provide any information about discontinuous displacements after the break. The GPU implementation of the hDIC technique with the new smoothing procedure was used for the identification of discontinuous displacements after the break of the same tensile test specimen. All calculations for the displacement calculations and benchmark tests were conducted with single precision.

3.1. Sample Preparation and Profilometry

Tensile test specimen of aluminium alloy 6082/HE30 was prepared using the electric discharge machining technique which creates an electric arc that causes formation of pits with a depth of a few microns on the cut surface. Elastic modulus and Poisson's ratio of HE 30 6082 aluminium are 70 GPa and 0.33, respectively. The tensile test specimen has a thickness of 2 mm and details of all dimensions can be found in the paper that presents the hDIC technique [15]. Tension load was applied using a 5kN tensile stage (Deben, Suffolk, UK) until the break and in situ surface profile scans were performed using an Infinite Focus 3D Profilometer instrument (Bruker Alicona, Graz, Austria). Vertical and lateral resolutions were kept as 200 nm and 8 μm, respectively, during the profilometry scans. Settings of the profilometer on the resolution was crucial, because it should be kept the same in order to satisfy repeatability. On the other hand, optical settings of the device were kept as default. Topography data corresponding to the reference and target conditions, as well as the information about the triaxial coordinates of the surface, are illustrated in Figure 2. The gap between the two sides after the break was minimised until the two edges got in touch, as illustrated in the circled images. The illustration in that figure shows that some parts of the broken edges are in touch. The top of the right side of the broken edge has a sharp jut along the top line while the symmetric line on the left side was buckled towards the centre of the material.

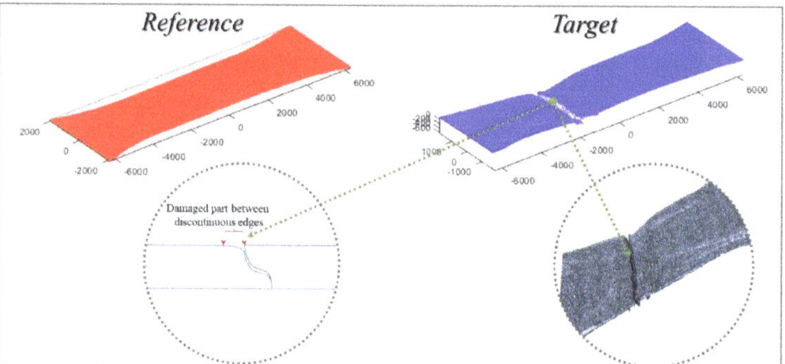

Figure 2. Topography data of the reference and the target conditions with illustration of the buckled part on the left side of the break zone.

3.2. Determination of Optimum Subset Size and Smoothness

GPU implementation of the hDIC technique with the new smoothing procedure has two parameters that need to be determined carefully, which are smoothness and subset size. These parameters were investigated in terms of the coefficient of variation (CoV) of the error between the pixel intensities of the reference and the target condition. Optimum parameters are selected from the ones that provide the minimum CoV of the error. Tests were conducted by running both the integer-pixel and sub-pixel level correlation stages for each parameter using the same region of interest (ROI) which spreads from −4 mm to 3 mm along the longitudinal direction in the break zone with respect to the centre of the gauge, and covers the gauge width with a range of 3 mm, as illustrated in Figure 3.

The first parameter is a subset size that should be determined carefully in order to calculate discontinuous displacements corresponding to the buckled and damaged sections of the break zone accurately, because the correlation on the damaged parts is expected to be accomplished by matching the pixels of the subset that belongs to the undamaged regions. The optimum subset size is determined by testing different subsets with a varying length of square, from 3 to 41 pixels, while keeping the smoothing parameter at maximum. Results of this analysis are given in Figure 4 in terms of the logarithm of the CoV. Minimum CoV is achieved when the subset length is 17 pixels, which corresponds to a subset of 289 pixels.

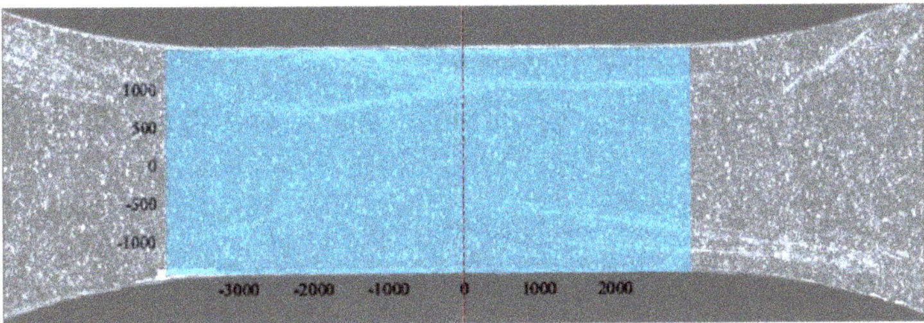

Figure 3. Microscopy image of the reference condition of the tensile test specimen and illustration of position and dimensions of the region of interest (ROI) in units of micron, obtained using an Alicona Infinite Focus 3D Profilometer instrument.

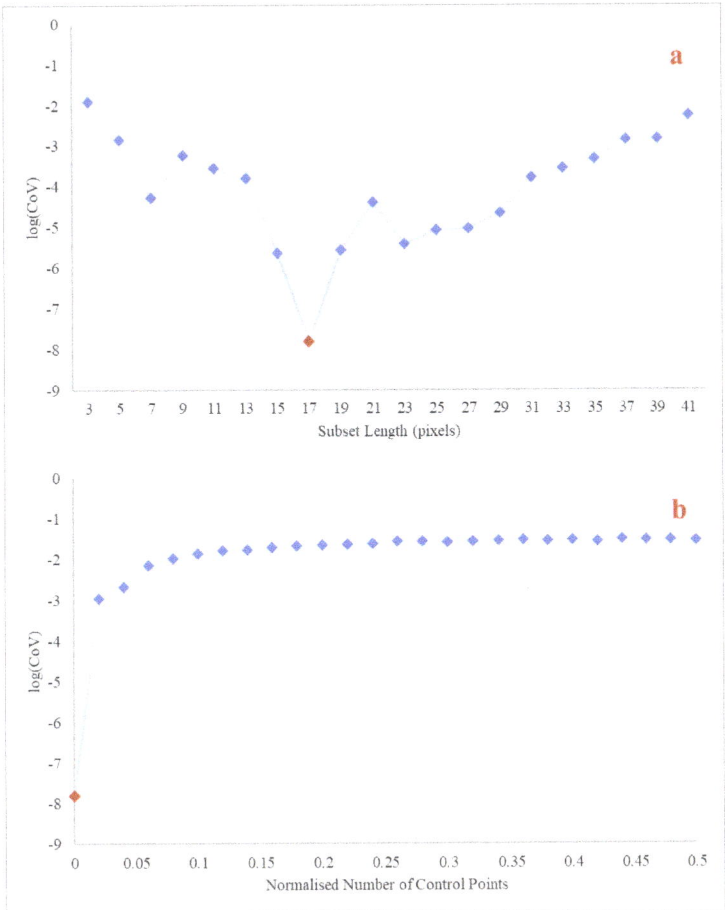

Figure 4. Variation of logarithm of the coefficient of variation (CoV) with respect to subset length (**a**) and normalised number of control points (**b**). There is a connection between (**a**) and (**b**) parts of this figure. The analysis on normalised number of control points is performed using the best subset length obtained after the analysis on subset length.

The second parameter is smoothness of the surface function that has crucial importance for cleaning the noisy displacement distribution and determination of the initial pixel for the iterations of sub-pixel level correlation stage. In this study, the purpose of the smoothing process is to create a function of smoothed results for the integer-pixel level correlation to predict coordinates of best matching pixels, but this smoothing process is not applied to the final results after the sub-pixel level correlation. The smoothness of the surface function created by the bi-cubic B-form least-squares spline interpolation [36] process is adjusted by the number of control points. Minimizing the number of control points increases the smoothness. In order to investigate the effect of smoothness on the CoV, smoothness is decreased to half, starting from the maximum smoothness while keeping the square subset length at 17 with 289 pixels. Results given in Figure 4 show that the minimum CoV is achieved when the smoothness is at maximum. However, this is achieved when the step size between subsets is one that means all pixels in the ROI were used in the correlation process.

3.3. Discontinuous Displacements after Break

After the parameter analysis of GPU implementation of the hDIC technique with the new smoothing procedure, optimum subset length and smoothness were used to determine discontinuous displacements on the tensile test specimen after break. Figure 5 illustrates smoothed displacements determined after the integer-pixel level cross-correlation stage and the sub-pixel level correlation stage displacement calculations without smoothing on the undeformed shape of the ROI. Results show that the range between maximum and minimum values of displacement were changed, clusters of pixels that have similar displacements appeared and discontinuous edges were visible after the correlation. As expected, the path of the discontinuous edge on the right side of the break zone appeared in the illustrations of x- and z-displacements and the continuous increase in z-displacement, up to the path of the discontinuous edge, was observed on the left side of the break zone. The path of the discontinuous edge on the right side of the break zone is less visible in the illustration of y-displacement.

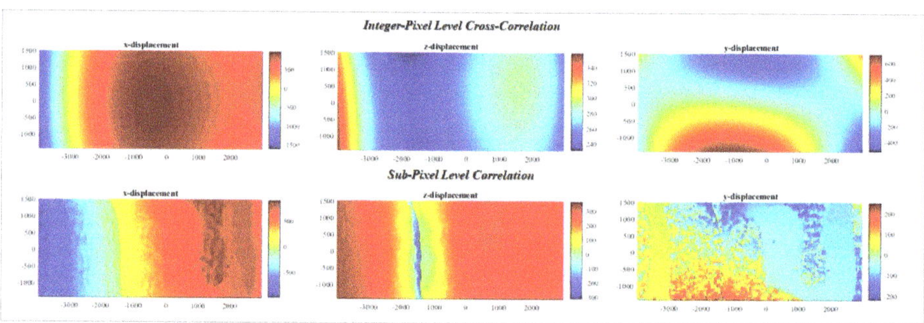

Figure 5. Results of the hDIC calculations after integer-pixel level cross-correlation and sub-pixel level correlation stages represented in the undeformed ROI in units of micron.

Displacements on the deformed shape of the ROI are illustrated as the transition from the microscopy image to the calculated displacements in Figure 6. This illustration shows the distribution of displacements on the deformed body calculated after the sub-pixel level correlation stage. Results of x- and z-displacements show a perfect match between the path of the discontinuous edge on the right side of the break zone in the microscopy image and the hDIC calculations. The discontinuous region becomes visible in the representation of all components of dispalcement but it becomes sharply visible in the z-displacement results seen in the middle column of Figure 6. This result shows the importance and benefits of using surface topography information for the purpose of the determination of discontinuous displacements.

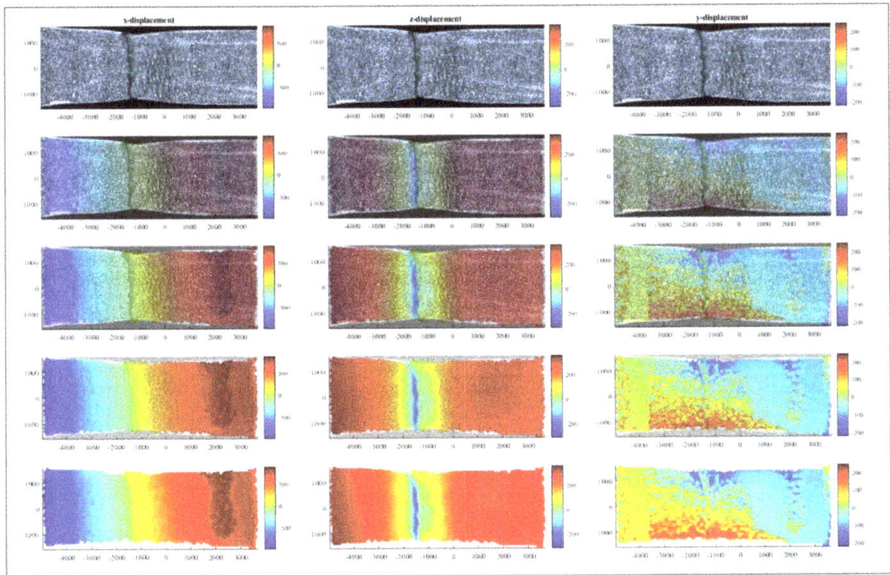

Figure 6. Illustration of the hDIC sub-pixel level displacement calculations in the deformed ROI and their transition from microscopy image of the deformed tensile test specimen in units of micron.

Local standard deviation around the break zone was calculated for better representation of discontinuous edges and illustrated in Figure 7. Coordinates on the microscopy image are the same with coordinates given in Figure 6 and in order to get a clear representation of the break zone, coordinates are not presented in this image. As expected, local standard deviations of x- and z-displacements provide a very clear view of the discontinuous edges on the right side of the break zone. The representation of local standard deviation of z-displacements also makes the beginning of the buckled field on the left side of the break zone visible, especially on the central parts of the gauge width, because in that region buckling is higher, but its distribution is not continuous.

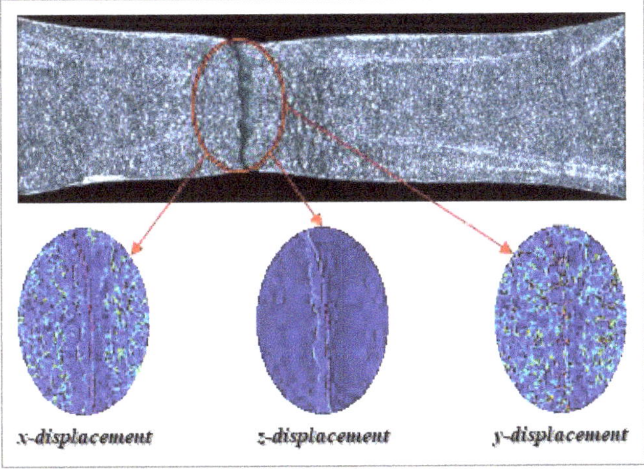

Figure 7. Illustration of local standard deviation of the hDIC displacement calculations around the break zone.

GPU parallel computing implementation of the hDIC technique was created based on Compute Unified Device Architecture (CUDA) parallel computing platform. In order to investigate the performance improvement gained by the GPU parallel computing, benchmark tests were conducted to compare serial calculations of Intel i7-8700 CPU (Intel, Santa Clara, CA, USA) and parallel calculations Nvidia Quadro P2000 GPU (Nvidia, Santa Clara, CA, USA). Tests were performed in ROIs with varying size and the results are illustrated in Figure 8. Normalised solution times show that the subpixel level correlation stage is completed five times faster in both CPU and GPU solutions which are linearly dependent on the number of pixels. The speedup gained by GPU parallel computing increases with the increasing number of pixels at each correlation stage. Maximum speedup is 6.2 times of the CPU calculations which was achieved when the ROI was created using the maximum number of pixels in the integer-pixel level cross-correlation stage.

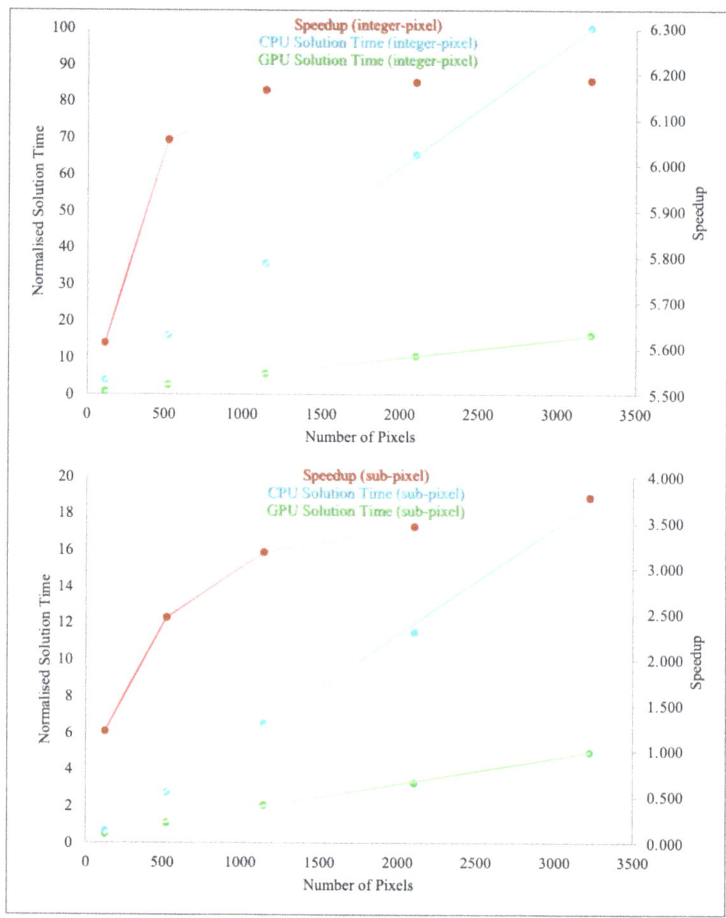

Figure 8. Results of the benchmark tests in terms of normalised solution times of serial CPU and parallel GPU computing.

Variation of the speedup in the sub-pixel level correlation stage shows that the increasing trend continues with increasing number of pixels, but the rate of the increase in the speedup decreases with increasing number of pixels. The speedup gained using this GPU has a potential to increase with increasing number of pixels in the sub-pixel level correlation stage, if the number of pixels is increased.

The expectation for the maximum speed up in the sub-pixel level correlation stage is to reach a closer value obtained in the integer-pixel level correlation stage.

4. Conclusions

The hDIC technique was initially developed for the purpose of the determination of continuous displacement fields in three axes at the surface of a material using surface topography [15]. The proposed solution procedure of the hDIC technique and the use of the optimised subset size and smoothness allowed for the identification of discontinuous displacements involuntarily using surface topography information without any need for user interaction and guiding data. The new form of the hDIC technique correlates the surface topography information of reference and damaged states successfully in the regions where the material was buckled or broken. Boundaries of the discontinuous edges become sharply visible in the displacement results. The performance of the proposed method was improved by GPU implementation. The use of the paging method allowed us to speedup integer-pixel and sub-pixel level correlation stages. Visual representations of discontinuous edge on the right side of the break zone showed perfect agreement with the illustrations in microscopy image. Further improvements can be achieved by increasing the density of topography data.

Author Contributions: Conceptualization, F.U. and A.M.K; methodology, F.U. and A.M.K.; software, F.U.; validation, F.U. and A.M.K.; formal analysis, F.U.; writing-original draft preparation, F.U.; writing-review and editing, F.U. and A.M.K.; visualization, F.U.; supervision, A.M.K.; project administration, F.U.; funding acquisition, F.U. All authors have read and agreed to the published version of the manuscript.

Funding: This project has received funding from the European Union's Horizon 2020 research and innovation programme under the Marie Skłodowska-Curie grant agreement No RESTREIG

Nomenclature

(x', y')	coordinates in the target subset
$B_{i,k}(x)$	i^{th} B-splines with a degree of $(k-1)$
$B_{j,k}(y)$	j^{th} B-splines with a degree of $(k-1)$
n_x, n_x	number of control points
R_m	denotes the mean intensity of the pixels in the reference subset
T_m	mean intensity of the pixels in the target subset
$T_s(x', y')$	intensity of a pixel at coordinates (x', y')
(x, y)	coordinates in the reference subset
$\alpha_{i,j}, \beta_{i,j}$	coefficients of the spline function
$\omega_{i,j}$	coefficient at the dimensions of i and j
i, j	dimensions of the spline function
N	an odd number to determine subset size
$R(x, y)$	intensity of a pixel at coordinates (x, y)
$T(x', y')$	intensity of a pixel at coordinates (x', y')
u	displacement in x-axis
v	displacement in y-axis
γ	step size of the gradient descent function

References

1. Parks, V.J. The range of speckle metrology. *Exp. Mech.* **1980**, *20*, 181–191. [CrossRef]
2. Chu, T.C.; Ranson, W.F.; Sutton, M.A. Applications of digital-image-correlation techniques to experimental mechanics. *Exp. Mech.* **1985**, *25*, 232–244. [CrossRef]
3. Hild, F.; Roux, S. Digital image correlation: From displacement measurement to identification of elastic properties—A review. *Strain* **2006**, *42*, 69–80. [CrossRef]
4. Zhao, P.; Zsaki, A.M.; Nokken, M.R. Using digital image correlation to evaluate plastic shrinkage cracking in cement-based materials. *Constr. Build. Mater.* **2018**, *182*, 108–117. [CrossRef]

5.   Li, X.; Xu, W.; Sutton, M.A.; Mello, M. Nanoscale deformation and cracking studies of advanced metal evaporated magnetic tapes using atomic force microscopy and digital image correlation techniques. *Mater. Sci. Technol.* **2006**, *22*, 835–844. [CrossRef]

6.   Sun, Y.; Pang, J.H.L.; Fan, W. Nanoscale deformation measurement of microscale interconnection assemblies by a digital image correlation technique. *Nanotechnology* **2007**, *18*. [CrossRef]

7.   Xu, Z.H.; Sutton, M.A.; Li, X. Mapping nanoscale wear field by combined atomic force microscopy and digital image correlation techniques. *Acta Mater.* **2008**, *56*, 6304–6309. [CrossRef]

8.   Bruno, L. Full-field measurement with nanometric accuracy of 3D superficial displacements by digital profile correlation: A powerful tool for mechanics of materials. *Mater. Des.* **2018**, *159*, 170–185. [CrossRef]

9.   Vendroux, G.; Knauss, W.G. Submicron Deformation Field Measurements: Part 2. Improved Digital Image Correlation. *Exp. Mech.* **1998**, *38*, 86–92. [CrossRef]

10.   Bertin, M.; Du, C.; Hoefnagels, J.P.M.; Hild, F. Crystal plasticity parameter identification with 3D measurements and Integrated Digital Image Correlation. *Acta Mater.* **2016**, *116*, 321–331. [CrossRef]

11.   Wang, B.; Pan, B. Subset-based local vs. finite element-based global digital image correlation: A comparison study. *Theor. Appl. Mech. Lett.* **2016**, *6*, 200–208. [CrossRef]

12.   Hu, Z.; Xie, H.; Lu, J.; Hua, T.; Zhu, J. Study of the performance of different subpixel image correlation methods in 3D digital image correlation. *Appl. Opt.* **2010**, *49*, 4044–4051. [CrossRef] [PubMed]

13.   Pan, B.; Yu, L.P.; Zhang, Q.B. Review of single-camera stereo-digital image correlation techniques for full-field 3D shape and deformation measurement. *Sci. China Technol. Sci.* **2018**, *61*, 2–20. [CrossRef]

14.   Prakoonwit, S.; Benjamin, R. 3D surface reconstruction from multiview photographic images using 2D edge contours. *3D Res.* **2012**, *3*, 1–12. [CrossRef]

15.   Uzun, F.; Korsunsky, A.M. The Height Digital Image Correlation (hDIC) Technique for the Identification of Triaxial Surface Deformations. *Int. J. Mech. Sci.* **2019**, *159*, 417–423. [CrossRef]

16.   van Beeck, J.; Neggers, J.; Schreurs, P.J.G.; Hoefnagels, J.P.M.; Geers, M.G.D. Quantification of Three-Dimensional Surface Deformation using Global Digital Image Correlation. *Exp. Mech.* **2014**, *54*, 557–570. [CrossRef]

17.   Sause, M.G.R. Digital image correlation. *Springer Ser. Mater. Sci.* **2016**, *242*, 57–129. [CrossRef]

18.   Abanto-Bueno, J.; Lambros, J. Investigation of crack growth in functionally graded materials using digital image correlation. *Eng. Fract. Mech.* **2002**, *69*, 1695–1711. [CrossRef]

19.   Mathieu, F.; Hild, F.; Roux, S. Identification of a crack propagation law by digital image correlation. *Int. J. Fatigue* **2012**, *36*, 146–154. [CrossRef]

20.   Mokhtarishirazabad, M.; Lopez-Crespo, P.; Moreno, B.; Lopez-Moreno, A.; Zanganeh, M. Evaluation of crack-tip fields from DIC data: A parametric study. *Int. J. Fatigue* **2015**, *89*, 11–19. [CrossRef]

21.   Hamam, R.; Hild, F.; Roux, S. Stress intensity factor gauging by digital image correlation: Application in cyclic fatigue. *Strain* **2007**, *43*, 181–192. [CrossRef]

22.   Mcneill, S.R.; Peters, W.H.; Sutton, M.A. Estimation of stress intensity factor by digital image correelation. *Eng. Fract. Mech.* **1987**, *28*, 101–112. [CrossRef]

23.   Fayyad, T.M.; Lees, J.M. Application of Digital Image Correlation to Reinforced Concrete Fracture. *Procedia Mater. Sci.* **2014**, *3*, 1585–1590. [CrossRef]

24.   Chernyatin, A.S.; Matvienko, Y.G.; Lopez-Crespo, P. Mathematical and numerical correction of the DIC displacements for determination of stress field along crack front. *Procedia Struct. Integr.* **2016**, *2*, 2650–2658. [CrossRef]

25.   Jandejsek, I.; Gajdoš, L.; Šperl, M.; Vavřík, D. Analysis of standard fracture toughness test based on digital image correlation data. *Eng. Fract. Mech.* **2017**, *182*, 607–620. [CrossRef]

26.   Hosdez, J.; Witz, J.F.; Martel, C.; Limodin, N.; Najjar, D.; Charkaluk, E.; Osmond, P.; Szmytka, F. Fatigue crack growth law identification by Digital Image Correlation and electrical potential method for ductile cast iron. *Eng. Fract. Mech.* **2017**, *182*, 577–594. [CrossRef]

27.   Bourdin, F.; Stinville, J.C.; Echlin, M.P.; Callahan, P.G.; Lenthe, W.C.; Torbet, C.J.; Texier, D.; Bridier, F.; Cormier, J.; Villechaise, P.; et al. Measurements of plastic localization by heaviside-digital image correlation. *Acta Mater.* **2018**, *157*, 307–325. [CrossRef]

28.   Cinar, A.F.; Barhli, S.M.; Hollis, D.; Flansbjer, M.; Tomlinson, R.A.; Marrow, T.J.; Mostafavi, M. An autonomous surface discontinuity detection and quantification method by digital image correlation and phase congruency. *Opt. Lasers Eng.* **2017**, *96*, 94–106. [CrossRef]

29. Uzun, F.; Salimon, A.I.; Statnik, E.S.; Besnard, C.; Chen, J.; Moxham, T.; Salvati, E.; Wang, Z.; Korsunsky, A.M. Polar transformation of 2D X-ray diffraction patterns and the experimental validation of the hDIC technique. *Measurement* **2019**, 107193. [CrossRef]

30. Craven, P.; Wahba, G. Smoothing noisy data with spline functions—Estimating the correct degree of smoothing by the method of generalized cross-validation. *Numer. Math.* **1978**, *31*, 377–403. [CrossRef]

31. Woltring, H.J. On optimal smoothing and derivative estimation from noisy displacement data in biomechanics. *Hum. Mov. Sci.* **1985**, *4*, 229–245. [CrossRef]

32. Palanca, M.; Tozzi, G.; Cristofolini, L. The use of digital image correlation in the biomechanical area: A review. *Int. Biomech.* **2016**, *3*, 1–21. [CrossRef]

33. Li, X.; Zhao, J.; Shuai, J.; Zhang, Z.; Wu, X. Accurate Reconstruction of High-Gradient Strain Field in Digital Image Correlation: A Local Hermite Scheme. In *Advancement of Optical Methods & Digital Image Correlation in Experimental Mechanics, Volume 3*; Lamberti, L., Lin, M.-T., Furlong, C., Sciammarella, C., Reu, P.L., Sutton, M.A., Eds.; Springer International Publishing: Berlin/Heidelberg, Germany, 2019; pp. 173–175.

34. Réthoré, J.; Gravouil, A.; Morestin, F.; Combescure, A. Estimation of mixed-mode stress intensity factors using digital image correlation and an interaction integral. *Int. J. Fract.* **2005**, *132*, 65–79. [CrossRef]

35. Russell, W.S. Polynomial interpolation schemes for internal derivative distributions on structured grids. *Appl. Numer. Math.* **1995**, *17*, 129–171. [CrossRef]

36. Amiri-Simkooei, A.R.; Hosseini-Asl, M.; Safari, A. Least squares 2D bi-cubic spline approximation: Theory and applications. *Meas. J. Int. Meas. Confed.* **2018**, *127*, 366–378. [CrossRef]

Article

High Cycle Fatigue Data Transferability of MAR-M 247 Superalloy from Separately Cast Specimens to Real Gas Turbine Blade

Miroslav Šmíd [1,2], Vít Horník [1], Ludvík Kunz [1], Karel Hrbáček [1,3] and Pavel Hutař [1,2,*]

1   Institute of Physics of Materials, Czech Academy of Sciences, Žižkova 22, 616 62 Brno, Czech Republic; smid@ipm.cz (M.Š.); hornik@ipm.cz (V.H.); kunz@ipm.cz (L.K.); Hrbacek.Karel@pbs.cz (K.H.)
2   CEITEC IPM, Žižkova 22, 616 62 Brno, Czech Republic
3   PBS Velká Bíteš a.s., Vlkovská 279, 595 12 Velká Bíteš, Czech Republic
*   Correspondence: hutar@ipm.cz; Tel.: +420-532-290-351

Received: 17 September 2020; Accepted: 29 October 2020; Published: 31 October 2020

Abstract: Cast polycrystalline superalloys are widely used for critical components in aerospace and automotive industries, such as turbine blades or turbocharges. Therefore, their fatigue endurance belongs to one of the most essential mechanical characteristics. Full-scale testing of such components involves great technical difficulties and requires significant experimental effort. The present study evaluates the effects of microstructural parameters with respect to representative fatigue testing of a cast turbine blade by separately cast specimens. For that purpose, the cast polycrystalline MAR-M 247 Ni-based superalloy was investigated in the following conditions: (i) specimens extracted from a real gas turbine blade; specimens separately cast into the mould with (ii) top or (iii) bottom filling systems. Obtained diverse microstructures allowed us to assess the effect of grain size, porosity, and texture on fatigue performance. The tests were held at a symmetrical loading regime at temperature 800 °C in laboratory air. The results indicate that the level of porosity is a dominant structural parameter determining the fatigue endurance, while grain size and texture effects were of minor importance contributing mainly to fatigue life scatter.

Keywords: high-temperature fatigue; nickel-based superalloy; investment casting; metallography; turbine blade

---

1. Introduction

Nickel-based superalloys are a vast class of the materials, essential for the aerospace, automotive, and power generation industry, which are recognized for an excellent combination of strength, structural stability at high temperatures, and very good corrosion resistance. Although single-crystal superalloys are used in modern gas turbines or jet engines, polycrystalline investment cast superalloy blades still find a wide array of applications thanks to less demanding fabrication processes and better affordability. Therefore, the investigation of fatigue performance and the effect of the structural imperfections naturally inherited by casting technologies still attracts interest [1–3].

Before the integration of a blade into a turbine, the assessment of design and fatigue endurance is essential and of great interest for aerospace and power generation industries. These components experience severe service conditions comprising mechanical loading (low and high cycle fatigue, creep), high temperatures, and corrosion environments, which may result in catastrophic failure. Since the geometrical complexity of the blades inevitably presents locally different casting conditions, such as temperature gradient and solidifications rate, the microstructure of various sections can differ significantly [4]. These variations are reflected by the diversity in grain size, primary and secondary dendritic arm spacing, distribution and size of carbides, shrinkages, and gas pores. Microstructural deviations result in locally dependent susceptibility to failure by different loading conditions within a blade. For instance, high cycle fatigue (HCF) damage processes are very localized and susceptible to local stress concentrators such as shrinkages and gas pores [5,6]. Contrary to that, creep and low cycle fatigue (LCF) damage mechanisms are affected by other microstructural aspects such as the distribution and the size of brittle phases [7,8], topologically close-packed phases, and γ/γ' eutectics [9]. Moreover, the microstructures with various phase constituents accommodate plastic deformation in a different manner reflected by different crack initiation and propagation mechanisms [2]. Therefore, the relationship of processing technology–microstructure–mechanical properties at severe service conditions possesses a high level of complexity and represents the important subject of an intensive effort for academia and industry.

Isothermal mechanical testing provides an essential experimental data for the determination of expected total operational lifetime of a turbine blade. Moreover, mechanical testing is essential for the maintenance interval optimization, which presents cost- and time-wise undesirable shut-downs. One of the possible approaches for fatigue and creep tests is the utilization of real full-scale blades. An undeniable advantage is that test blades undergo the identical manufacturing processes and possess the same microstructure and dimensions as the blade utilized in a real turbine. However, full-scale experiments are financially very demanding, time-consuming, and require a unique laboratory facility [1,10,11]. Another setback of such an approach is the inability to isolate the individual effects of various mechanical loadings, which makes the experimental data analysis difficult. Therefore, the finite element (FE) and crystal plasticity modelling [12] of turbine blades under various loading conditions is regarded as an important supplementary approach to analyse such complex engineering challenges [1,13–15]. Experimental results carried out on small-scale specimens extracted from various blade sections can be utilized as a data input or a verification basis of the models. This approach can bring a better understanding of mechanical property variations with respect to local microstructure and texture heterogeneity stemming from different solidification conditions [16–18]. A few small-scale specimen studies [19,20] were carried out in the past and have shown their worthiness by presenting significant dependence of the fatigue endurance, hardness, and tensile properties on a blade section.

The present study aims to highlight the important factors of high-temperature fatigue testing within the framework of the HCF performance assessment of a gas turbine blade. The fatigue tests were carried out using fatigue specimens produced by two casting processes varying in different mould filling setups. Acquired microstructures and related fatigue life are discussed and compared with the fatigue performance of the specimens extracted from three locations of a gas turbine blade.

2. Materials and Methods

The cast polycrystalline MAR-M 247 superalloy, provided by the company PBS Velká Bíteš, a.s., was processed by investment casting in three variants: (i) real blade casting, (ii) casting of semi-finished specimens by a mould with top filling (further entitled as TF), and (iii) with bottom filling (BF). Figure 1 shows the schematics with the highlighted flow of the molten metal during the casting. Different moulds were utilized in order to obtain various microstructures and to minimize porosity: eight cylindrical bars in an approximate shape of fatigue test specimens were attached to the top (Figure 1b) or the bottom radial running system (Figure 1c). The BF mould also contained the TF system, which was left just as a reinforcement to ease the handling. The casting was performed under vacuum 10^{-4} Pa, whereas pouring temperature was $1385 \pm 15\ °C$ and mould temperature was $1050\ °C$. After the end of solidification, the bars were extracted from the pattern and subsequently underwent hot isostatic pressing (HIP) procedure ($1200\ °C/4\ h/100\ MPa/argon$) and two-step heat treatment consisting of a solution annealing at $1200\ °C/2\ h/air$ cooling followed by a precipitation annealing at $870\ °C/24\ h/air$ cooling. Several bars cast by the TF mould were left without HIP to assess the effect of this process on the porosity and resulting fatigue life. Subsequently, the semi-finished specimens were machined into final specimen geometry.

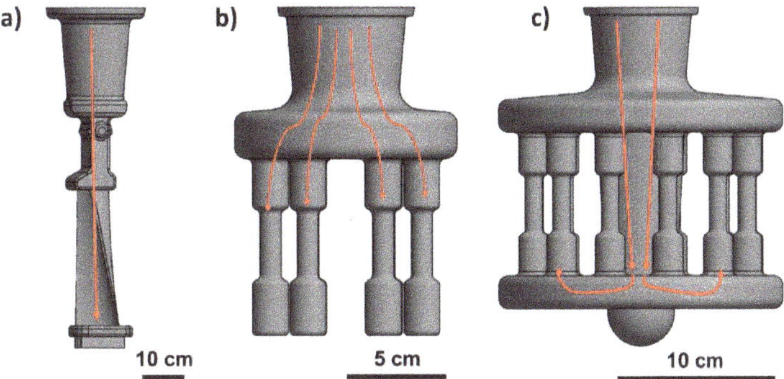

Figure 1. Schemes of the utilized casting setups: (**a**) real blade cast by the top filling system, (**b**) the mould with the top radial filling system, and (**c**) the pattern with the bottom radial filling system. The melt flow during pouring is depicted by arrows.

Besides the mould castings, the cylindrical bars dedicated for fatigue specimen fabrication were extracted by electrical discharge machining from three sections of the blade—further referred to as root, shank, and airfoil. From the reason of insufficient cross-blade dimensions, it was not possible to acquire specimens from the airfoil above cross-section III. The positions of bar extraction are depicted in Figure 2a and specimen geometry in Figure 2b. Due to complex geometry, the microstructure of the blade varied significantly. To examine this microstructural diversity, large-scale metallographic samples were prepared. The positions of sectioning, labelled I–IV, are depicted in Figure 2a, and the sample micrographs are shown in Figure S1 (available in the Supplementary Materials).

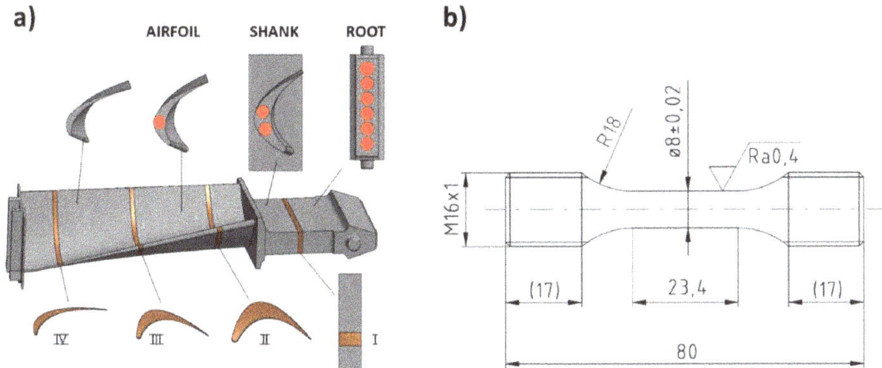

Figure 2. Schemes of (**a**) the blade cutting plan with indicated positions of extracted fatigue specimens (highlighted as red circles) and metallographic sample sections; (**b**) the specimen geometry.

A resonant testing machine Amsler with 100 kN force range was used for fatigue tests under load control regime in fully reversed loading ($R = -1$). The fatigue tests were performed at frequency 121 ± 2 Hz in laboratory air. The experimental temperature 800 °C was achieved by an electric furnace with resistance heating. The tests were terminated, while the loading frequency decreased by 5 Hz or in the case of specimen rupture. Fatigue limit, σ_f, was defined as a stress amplitude value at which a minimum of three specimens reached 2×10^7 cycles without failure.

The metallographic sections of blade, perpendicular to blade axis, and the sections of fatigued specimens, cut longitudinal to the loading direction, were carefully ground by sandpapers with increasing grit number and subsequently polished by 1 µm diamond paste. The final chemical–mechanical polishing by colloidal silica slurry provided the surface quality sufficient for the electron backscatter diffraction (EBSD) measurement. Etching by a solution of 100 mL H_2O, 4 mL HNO_3, and 2 mL HF was used to reveal grain boundaries. Initial light microscopic (LM) analysis was performed using a stereomicroscope Leica S9i, whereas fracture surfaces and longitudinal sections were observed by a scanning electron microscope (SEM) Tescan LYRA 3 XMU (Tescan Orsay Holding, a.s., Brno, Czech Republic). The EBSD observations were performed with an electron beam operating at 20 kV by a NordlysNano EBSD camera (Oxford Instruments, Abingdon, UK). The EBSD data were analysed by the CHANNEL 5 software (Oxford Instruments, Abingdon, UK). Secondary dendritic arm spacing (SDAS) measurement were performed by light microscope Olympus GX51 (Olympus, Tokyo, Japan) on polished and etched metallographic samples. For that purpose, the linear intercept method was employed, and presented values are the result of 60 measurements for each alloy variant.

Nominal chemical composition of the MAR-M 247 superalloy is given in Table 1. The alloy has a typical cast structure with coarse dendritic grains of average size ranging between hundreds of microns and a few millimetres. The microstructure, see Figure 3, contains strengthening coherent precipitates of γ' phase heterogeneously distributed in γ matrix. Areas of fine γ' phase with cuboidal shape (~0.4 μm) and areas of coarse γ' phase with more complicated morphology (~1.6 μm in diameter) were present. The investigations of Harris et al. [21] and Janowski [22] revealed the volume fraction of γ' phase of approximately 60%. Numerous carbides and γ/γ' eutectics were located along grain boundaries and interdendritic areas.

Table 1. Chemical composition of MAR–M 247 superalloy (in wt. %).

C	Cr	W	Co	Al	Ti	Ta	Hf	Mo	Nb	B	Ni
0.15	8.37	9.92	9.91	5.42	1.01	3.05	1.37	0.67	0.04	0.015	bal.

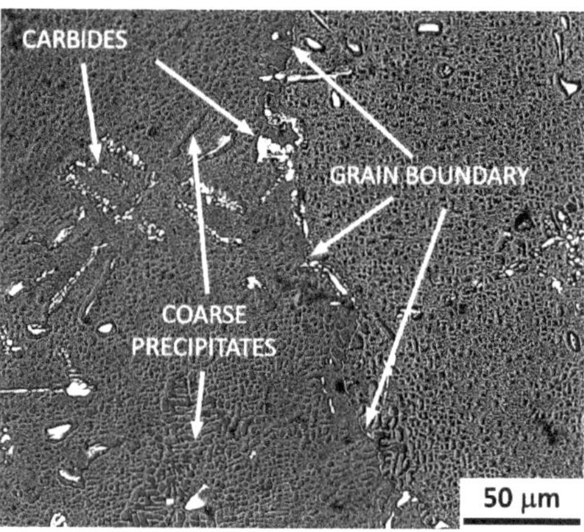

Figure 3. Microstructure of MAR-M 247 superalloy.

3. Results

3.1. Microstructural Characterization

Microstructures in the gauge part of the specimens extracted from all three blade positions are shown in Figure 4. Corresponding average grain size and secondary dendrite arm spacing (SDAS) are given in Table 2. Since grain size strongly varies, minimum and maximum values are also included. The significant scatter of grain size among the blade positions stems from the variations in local blade thickness and geometry. The root of the blade consists of large equiaxed grains owing to bulky character of the section resulting from slow solidification of the melt. The shank section consisted of distinctly finer dendritic structure. Further grain refinement was observed in the lower part of the airfoil. Since the specimens were fabricated from rather thick parts of the blade, equiaxed grains were dominant over columnar grains. In summary, distinct grain size refinement in the direction of the blade axis towards the blade tip is associated with the local blade thickness. This fact is also reflected by the decrease in the SDAS parameter, which is, together with the primary dendritic arms spacing, proportionally related to the temperature gradient G and the growth rate of the solidification front v represented by the following equation [23]:

$$\lambda_1 \approx v^{-1/4}\, G^{-1/2}. \tag{1}$$

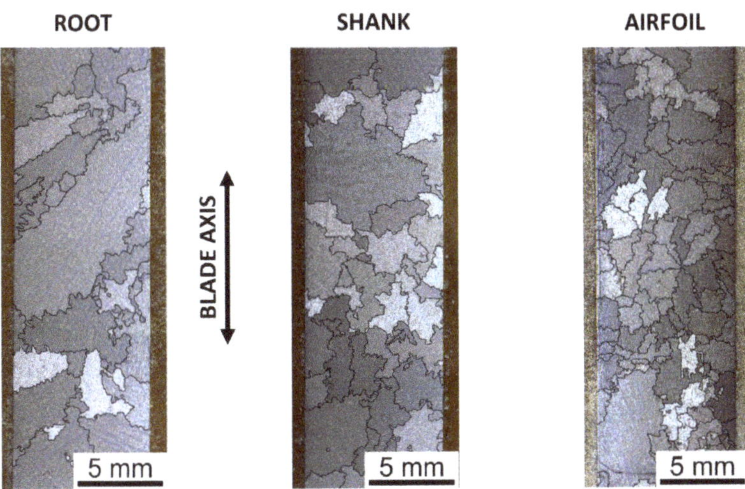

Figure 4. Structure of the specimen gauge sections extracted from three different positions of a gas turbine blade.

Table 2. Basic microstructural parameters.

	Average Grain Area [mm^2]	Grain Size Scatter [mm^2]	SDAS [µm]
Root	4.91	0.1–51.21	136 ± 39
Shank	4.35	0.02–30.62	105 ± 24
Airfoil	1.46	0.01–19.36	101 ± 27
Top filled mould without HIP *	0.33	0.01–5.76	33 ± 5
Top filled mould with HIP	0.36	0.02–6.14	34 ± 8
Bottom filled mould	2.45	0.02–18.66	39 ± 8

* Hot isostatic pressing (HIP).

Therefore, it can be assumed that the bulky blade positions solidify by lower rates, resulting in coarse grain structure with large SDAS, whereas the slender parts of the blade airfoil undergo faster solidification leading to finer dendritic microstructure and lower SDAS.

Figure 5 shows the structures of the TF mould cast specimens (TF specimens) without or with subsequent HIP processing and the BF mould cast specimens (BF specimens). Microstructural characterization was performed on metallographic samples from the specimen gauge cut parallel to the specimen axis. Utilized mould filling setups resulted in different structures in terms of grain size, grain columnarity, and porosity. As shown in Table 2, the average grain size of the BF specimens was notably larger as well as grain size scatter and SDAS parameter. These findings distinctly indicate that BF casting microstructure stems from the slower solidification due to the additional bottom radial running system, which adds a significant molten metal mass. Columnar grain structure was found within the TF specimens. Any distinct effect of the HIP on grain size was not found, which complies well with available literature [24,25].

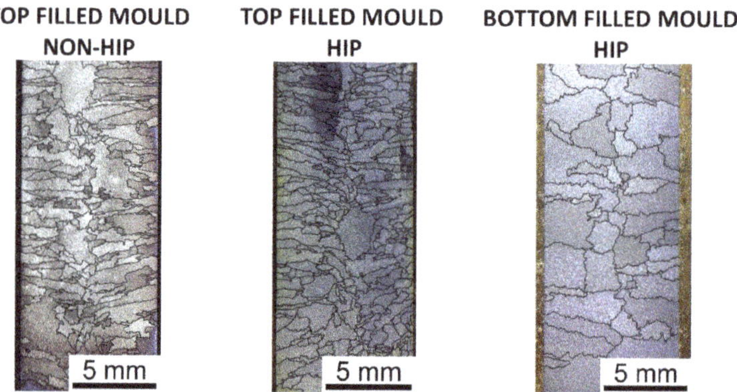

Figure 5. Dendritic structures of the specimen gauge sections.

The porosity assessment was carried out on the fracture surfaces of all fatigued specimens based on the assumption that the primary fatigue crack originates from the most serious structural defect of the specimen gauge. This assumption is deemed to be generally valid for HCF loading regime known for high sensitivity on structural imperfections and essential effect on the fatigue crack initiation [26,27]. The examination of fracture surfaces revealed that the BF specimens exhibited low porosity level with isolated pores in size range 30–70 µm, whereas the TF mould casting resulted in the microstructure containing the shrinkages with the average size 770 µm. The effect of the HIP process, evaluated on a series of the TF specimens left without this treatment, was found to be essential. Shrinkages and pore clusters of average size 3040 µm were found. The analysis of the blades in each position revealed fine pores; therefore, it can be concluded that, with respect to typical grain size observed in each of the blade sections, the porosity was negligible. The principal defects of each tested specimen characterized by Feret diameter are shown further in Figure 8 of Section 3.2.

The texture analysis of the blade shank section and both filling systems are shown in Figure 6. The pole figures acquired by the EBSD measurement were performed over a large area of the specimen gauge section depicted in Figure 5. High intensities within corresponding {001} pole figures were caused by coarse microstructures where individual grains cover a significant area fraction of an EBSD map. This is especially distinct in the case of the blade shank, which possessed the largest grain size within the examined microstructures. Despite this fact, any preferential grain orientation was not found.

In summary, the microstructures of specimens cast by three casting setups resulted in distinct differences in grain size and porosity, while the random texture of grains was retained.

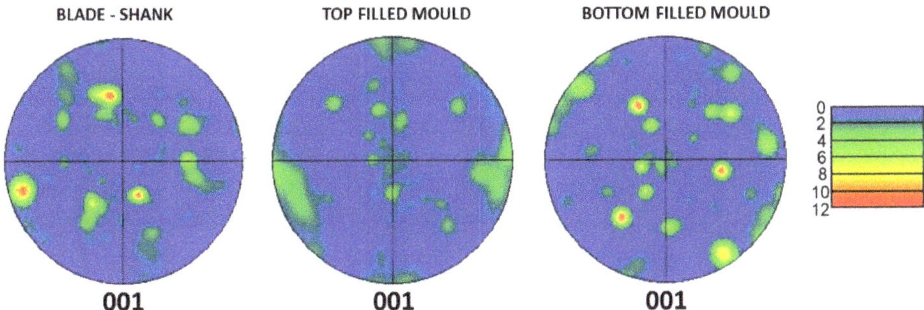

Figure 6. (001) pole of longitudinal specimen gauge sections obtained by three different casting conditions depict random crystallographic orientation. High intensities are caused by coarse grains.

3.2. Fatigue Life

Stress-life (S-N) data from the fatigue tests carried out on the specimens originating from three blade positions are shown in Figure 7. Despite distinct microstructural variation among individual positions (see Figure 4), the results indicate a negligible effect of the grain size on HCF performance. Therefore, all experimental data are fitted by a single fatigue life curve, and a fatigue limit of 230 MPa was identified for the blade regardless of the specimen position. In contrast to the similarities of the HCF performance, the scatter of fatigue data varies among the blade positions. Figure 7 indicates that the highest scatter is observed for the airfoil, the section possessing the finest microstructure. It is important to note that a limited number of tests was performed for the airfoil section, since only one specimen could be extracted per blade. Therefore, the experimental data for this particular section are not fully statistically sound in terms of fatigue performance assessment but can give sufficient insight into observed significant fatigue life scatter. Microstructural examinations of several specimen gauge sections revealed that the airfoil part of a blade is sensitive on inevitable fluctuations in casting conditions and in the solidification process. The occasional occurrence of individual coarse grains was confirmed in several specimens. Such microstructural variations can be the source of this fatigue data scatter.

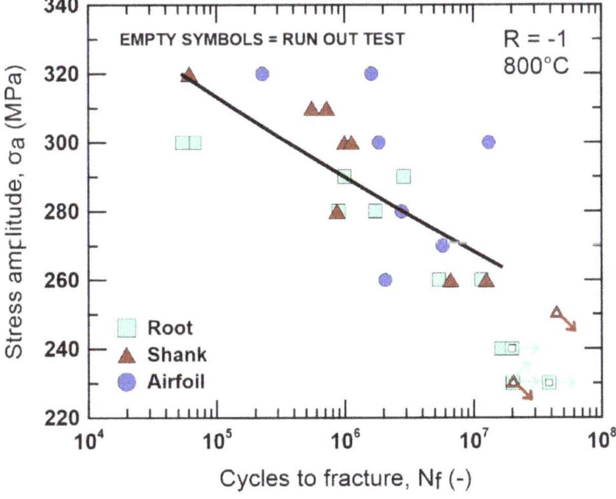

Figure 7. Stress-life (S-N) diagram for individual blade positions cycled at 800 °C.

The fatigue life curves of the specimens cast into the TF and the BF moulds are plotted in Figure 8. The diagram includes Feret diameter measurements of the casting defects detected on the fracture surface of each specimen and which were responsible for the fatigue crack initiation. The specimens with no apparent pores or shrinkages to trigger the fatigue crack initiation are the data points without number, which was a domain for the BF specimens. Metallographic samples and fracture surfaces of this casting variant revealed a low level of porosity with isolated gas pores and rare shrinkages. SEM fractographs of typical fracture surfaces for specimen variants is shown in Figure S2 (available in Supplementary Materials). Corresponding fatigue life curves show distinct differences, which directly correspond with the degree of porosity. The castings with a lower content of microstructural defects exhibit better HCF performance. The results reflect the positive effect of the BF pattern casting. Even larger influence than type of the filling system can be attributed to the HIP treatment, which resulted in fatigue life enhancement by more than two orders. Contrarily, the structure without HIP manifests the HCF performance of the alloy with initial shrinkages and gas pores created during solidification. With respect to the Feret diameter of the casting defects, the gauge area cross section is severely reduced leading to increased local peak stresses compared to applied macroscopic stress. Therefore, the fatigue crack initiation within favourably oriented grains adjacent to a large shrinkage is significantly accelerated as well as crack propagation, displayed clearly by strong fatigue life reduction. Fatigue limits were measured just for the HIPed alloy variants. The BF mould casting provides the fatigue limit by 30 MPa higher compared to 190 MPa of the TF mould casting. In summary, Figure 8 demonstrates the dominant effect of the porosity on HCF performance.

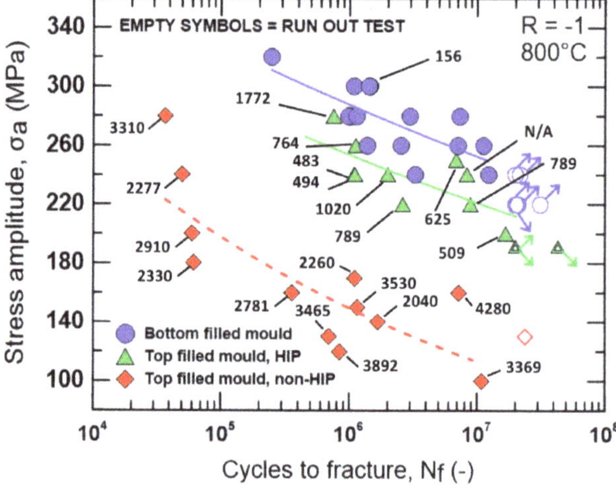

Figure 8. S-N diagram of particular batches of pre-cast specimens cycled at 800 °C. The numbers of each data point represent the Feret diameter of a casting defect (in μm) responsible for the fatigue crack initiation.

3.3. Fractographic Analysis

Figures 9–11 show micrograph compilations of selected failed specimens representing each casting variant. A fracture surface with a detail of the crack initiation site and an EBSD map of the grains in the fracture surface vicinity are depicted. The EBSD analysis was carried out on longitudinal sections of specimen gauge, which were cut to intersect the crack initiation site. The sectioning is highlighted by a yellow dashed line in each figure.

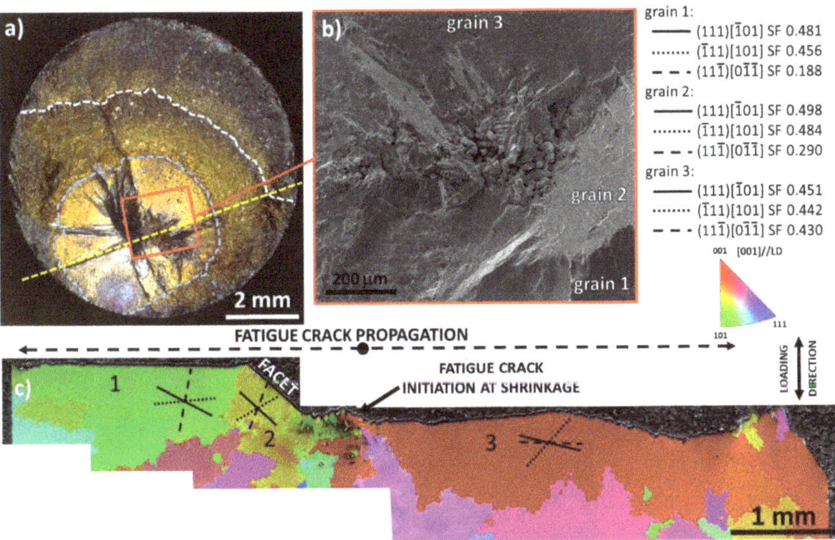

Figure 9. Fracture surface of top filling (TF) hot isostatic pressing (HIP)ed specimen (σ_a = 280 MPa, N_f = 0.752 × 10^6 cycles) (**a**). Fatigue crack initiated from large complex shrinkage situated in gauge length interior (**b**). The blue dashed line highlights the "fish-eye" representing the fatigue crack propagation period prior to reaching a specimen surface. The white dashed line refers to the final stage of crack propagation before final failure. The yellow line represents the plane of longitudinal section for electron backscatter diffraction (EBSD) analysis underneath the surface (**c**).

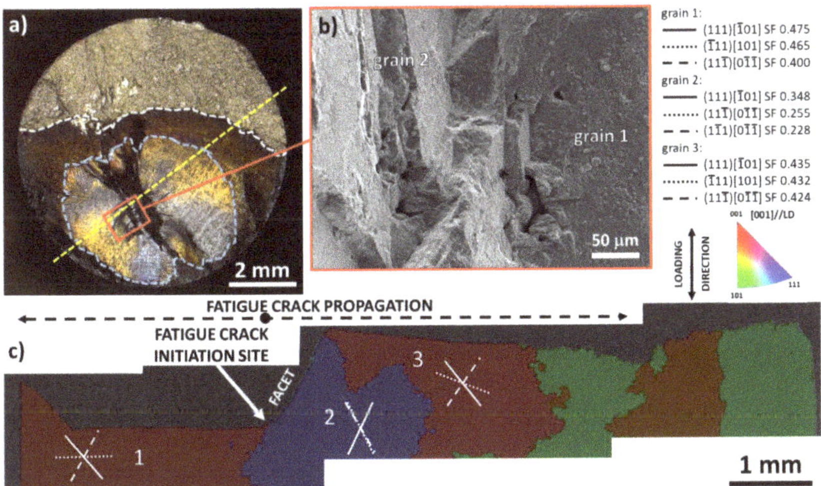

Figure 10. Fracture surface of bottom filling (BF) specimen (σ_a = 300 MPa, N_f = 1.438 × 10^6 cycles) (**a**). Fatigue crack initiated from small shrinkage (**b**). The blue dashed line highlights the fish-eye area representing fatigue crack propagation prior to reaching specimen surface. The white dashed line depicts the final stage of the crack propagation before final failure. The yellow line represents the plane of longitudinal section for EBSD analysis underneath the surface (**c**).

Figure 11. Fracture surface of blade shank specimen (σ_a = 300 MPa, N_f = 10^6 cycles) with surface crack initiation highlighted by a red rectangle (**a**). The white dashed line highlights final stage of the crack propagation before final failure. Detail of the crack initiation site is depicted in (**b**). The yellow line represents the plane of longitudinal section for EBSD analysis underneath the surface (**c**).

Figure 9 depicts the fracture surface of the TF HIP treated specimen. LM fractography, Figure 9a, revealed complex shrinkage in the sample interior identified as the crack initiation site. Large facets belonging to a single grain, namely grain 2, was found in the vicinity of the crack initiation site shown in Figure 9b. The early stage of the fatigue crack propagation without oxidation assistance is clearly visible as an area with markedly different fracture surface appearance, so-called "fish-eye", highlighted by the light blue dashed line. The further crack propagation until failure (white dashed line highlight) was characterized by pronounced fracture surface roughness closely related to the dendritic character of the microstructure. Carbide particles, γ/γ' eutectics and small pores were found on the fracture surface in this area indicating interdendritic crack propagation, see Figure S3 (available in Supplementary Materials). Slip trace analysis, Figure 9c, confirmed that the facet in the place of fatigue crack initiation was of a crystallographic nature, namely the slip plane of type {111} with the highest Schmid factor (SF). The cyclic loading forced the crack initiation along the slip plane with subsequent crack propagation in crystallographic stage I mode until the grain boundary with grain 1 was reached. Therefore, it can be assumed that the stress concentrations induced by the complex shrinkage promoted slip activity along the most suitably oriented slip plane of grain 2. The cyclic loading forced the crack initiation along the slip plane with subsequent crack propagation in crystallographic stage I mode until the grain boundary with grain 1 was reached. Afterwards, the crack driving energy was high enough to propagate further by non-crystallographic stage II mode via structurally weak spots such as carbides and γ/γ' eutectics. The slip trace analysis of grains 1 and 3 confirmed that the crack did not follow the high SF slip planes, i.e., non-crystallographic crack propagation was active.

As noted in the previous section, the porosity level dropped significantly by using the BF casting setup. As a result, the crack initiation was frequently observed along the favourably oriented slip systems of large grains, without distinct effect of porosity, and from the specimen surface. However, Figure 10 depicts fractographic analysis of the only crack initiation from the pore agglomeration documented in the BF mould castings. Due to suitable grain 1 orientation, the pores were capable of introducing cyclic slip activity on the slip plane with the highest SF. The crack initiation and early crack propagation were under crystallographic stage I mode, while the character of the fracture surface in grains 1 and 3 indicate a transition to non-crystallographic stage II mode. The fish-eye feature is again apparent on the fracture surface indicating oxidation unassisted crack propagation.

The notably different appearance of the fracture surface is presented for the failed specimen of the blade shank part. Since the porosity was nearly absent in the blade castings, the surface crack initiation was predominantly observed. The small facet within the grain 1, shown in Figure 11b, was identified as the fatigue crack initiation site. The crystallographic crack propagation was active only in grain 1. The fracture surface roughness, Figure 11c, gradually increased with the crack length in the same manner for TF and BF specimens. That indicates that the crack path led via fatigue weak areas such as interdendritic areas, as it was already documented by Figure S3 (available in Supplementary Materials).

4. Discussion

Small-scale fatigue specimens testing with the aim to characterize the fatigue performance of a real blade has been studied in the works of Yan et al. [20,28]. Attention was predominantly focused on the influence of specific blade geometry features and the local stress gradients on HCF and LCF life. However microstructural variability was not considered. The present study is focused on the microstructural characterization enabling subsequent evaluation of the dominant structural parameters affecting HCF, such as grain size, texture, and porosity. For that purpose, the fatigue specimens were prepared by two different mould filling setups in order to obtain diverse microstructures. In parallel, the additional set of specimens was extracted from the gas turbine blade in three positions, as it has been presented in Section 2.

The fatigue performance of the material extracted from all blade positions (shown in Figure 7) was similar; however, the fatigue life scatter exhibits variations, especially in the airfoil. The microstructure of this blade position consisted of finer dendritic grains with the occasional appearance of individual coarse grain. Since the level of porosity was negligible in the blade castings, coarse dendritic grains act as a fatigue hot spots where the cyclic deformation mechanisms were facilitated and subsequently a fatigue crack initiation can occur, as it has been described by Miao et al. [29]. The planar character of the cyclic plasticity, typical for superalloys even at a wide range of temperatures as shown by Šulák et al. [30], benefits from coarse grains in terms of dislocation slip activity along feasible crystallographic planes over long distances, uninterrupted by grain boundaries. Moreover, the damage mechanisms can be further accelerated by the occurrence of stress concentrators such as carbides and fine gas pores [31]. For instance, Du et al. [32] demonstrated the synergistic detrimental effect of structural casting defects and coarse grains, and their location with respect to the specimen surface. Furthermore, the fatigue crack growth rate is higher in coarse grain structure due to the lack of grain boundaries acting as a crack inhibitor [33,34]. Thus, the fatigue data scatter is attributed to the occurrence of coarse grains in some of the specimens. This conclusion implies that grain size variation, in the range found in this study, can contribute to increased fatigue data scatter. To significantly alter the fatigue lifetime, it is assumed that the grain size change would have to be in a considerably larger magnitude, as it was demonstrated by study of Du et al. [32].

It has been proposed by several studies [35,36] in the past that the crack initiation and early propagation in nickel-based superalloys occur by a slip decohesion mechanism with at least two active planes. This distinctly crystallographic mechanism implies that grain orientation will play an important role. Moreover, Liu et al. [34] have reported that grains with <001> direction parallel to the loading axis exhibit rapid crack propagation. The fracture surfaces, shown in the present work, possess distinct features of the crystallographic cracking in the form of facets; however, studied microstructures do not possess any preferential orientation, as shown by pole figures in Figure 6. The authors believe that the necessity of favourable grain orientation for fulfilling the condition of multiple active slip systems are bypassed by the local effect of defects in terms of induced stress concentrations. In such case, even secondary slip systems with lower SF will be activated. Therefore, the grain orientation will presumably play a less important role as the fatigue life determining microstructural parameter in comparison with porosity.

The fatigue life curves of the specimens cast by two different filling systems, shown in Figure 8, manifest the effect of casting defects, such as gas pores and shrinkages, on HCF performance. In general, any of the casting processes are commonly accompanied by some degree of porosity [37] and present the major factor for mechanical properties, mainly fatigue performance. The occurrence of a gas pore or a shrinkage depends on (i) the partial pressure of gas dissolved in a melt, (ii) the volume reduction during solidification, and (iii) the casting conditions [38]. Therefore, various types of filling systems will result in different casting soundness. It is generally regarded that the BF setup leads to less turbulent melt flow and consequently in a higher quality of the final component [39]. The mould with TF setup represents technologically easier process, but the melt flow is abrupt and turbulent and, therefore, prone to higher porosity and generally lower quality of a cast, as it was documented in the past [40]. The porosity measurement from the fracture surfaces for each filling system, shown in Figure 8, is in good agreement with these findings. Murakami [41] has proposed that material defects generate a stress intensity factor dependent on given defect size. The shrinkage-induced stress intensity factor introduces local stress concentrations leading to increased dislocation slip activity on some of the octahedral {111} planes. Such localization of cyclic plastic deformation can trigger the crack initiation in the vicinity of a defect followed by the crack propagation, which has been experimentally confirmed by digital image correlation investigations [31] or by EBSD [2]. In summary, a casting defect size, inherently induced by selected casting setup, has serious implications on the HCF performance of an alloy, which is reflected by presented fatigue data and has been confirmed also in different Ni-based superalloys [5], Ti-alloys [42], Al-alloys [43,44], and others [45]. The effect of porosity is further showcased by a set of specimens cast by TF setup without subsequent HIP treatment. Such a cast structure contained enormous shrinkages leading to accelerated fatigue damage process, which resulted in shorter fatigue life. Shrinkages of such extent led to a significant reduction in the specimen nominal cross-section, which increased applied stresses and intensify the stress concentrations. As a consequence, the fatigue crack initiation and propagation are accelerated. However, it has to be noted that the results contain several inconsistencies of the specimen containing a large shrinkage, which exhibited longer fatigue life than the specimen with smaller critical casting defects. Authors ascribe as a main source of such disparity: (i) relative position of a shrinkage with respect to the specimen surface and (ii) the character of microstructure in vicinity such as crystallographic orientation of adjacent grains.

Figure 12 depicts the concluding comparison of the fatigue data acquired by testing of specimens produced by both mould filling setups and the specimens extracted from the blade. Due to the comparable porosity level, the BF pattern specimens represent well the fatigue performance of the blade, supported by similar fatigue limit.

The fracture surfaces of all batches of specimens exhibited similar features. The vicinity of initiation sites frequently contains facets indicating crystallographic stage I mode of the fatigue crack growth. Facets are a direct consequence of intensive strain localization into favourably oriented slip planes, which are promoted by stress concentrators such as pores, shrinkages, carbides, and γ/γ' eutectics. In cast alloys, it is generally accepted that shrinkages and pores are the most critical structural defects [42,46]; therefore, the position of a fatigue crack initiation site is closely related to a position of casting defect, as it was already shown in the previous study [6]. In the case of a defect-free structure, the fatigue crack initiation from the fracture surface is the most frequent. This fact correlates well with the porosity measurements of the fatigued specimens presented in Figure 8 and the fracture analysis, shown in Figures 9–11 and S2. The turbine blade specimens or the BF specimens exhibited the negligible level of porosity resulting in the transition of the fatigue crack initiation on specimen surface where a contribution to the crack initiation stemming from hot corrosion can be expected. Contrary to this finding, TF specimens (with or without HIP) contain a considerable degree of casting defects especially in specimen interior which result in interior fatigue crack initiation and subsequent propagation. With further crack growth, the fracture surface becomes more rugged and a fatigue crack propagates via interdendritic areas containing a high amount of carbides and γ/γ' eutectics, see Figure S3.

The results confirm the dominant effect of casting defects on fatigue life, whereas grain size and texture are of minor effect. The separate casting of specimens for fatigue tests, aiming to avoid complex full-scale testing of blades or tedious and time-demanding extraction of test specimens from a blade, can deliver representative fatigue data under the condition of a similar level of porosity. Such a method of the fatigue tests can provide rapid and representative fatigue data for the design and manufacturing of a blade.

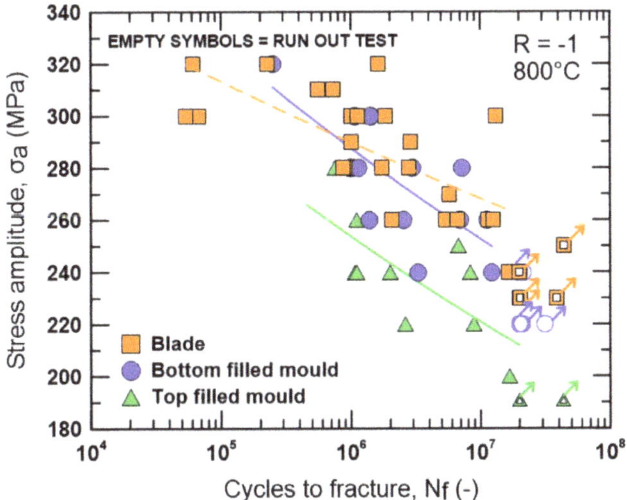

Figure 12. Comparison of high cycle fatigue (HCF) performance of the blade and the specimens cast by two different mould filling setups.

5. Conclusions

The results of this study indicate that the HCF life of polycrystalline cast MAR-M 247 superalloys at 800 °C is closely related to the retained porosity after the casting process and the HIP treatment. This finding presents a possibility for representative HCF testing of turbine blades by a set of separately cast specimens using standard laboratory fatigue testing systems. It has been concluded, based on the tests of three different series of specimens with diverse microstructural characteristics, that the determining parameter is the porosity size. The grain size and texture show only a minor effect on the overall HCF endurance and contribute most to the fatigue life scatter.

Supplementary Materials: The following are available online at http://www.mdpi.com/2075-4701/10/11/1460/s1, Figure S1: Macrostructure of the alloy in selected sections of the blade labelled according to the scheme shown in Figure 2a, Figure S2: Fractographs of (a) TF specimens without HIP, (b) TF specimen with HIP and (c) BF specimen, Figure S3: Backscattered electron micrograph of the fracture surface showing the presence of carbides and gas pores, favourable sites for the fatigue crack propagation.

Author Contributions: Conceptualization, M.Š., V.H. and P.H.; methodology, validation, investigation, data curation and visualization, M.Š. and V.H.; formal analysis and writing—original draft preparation, M.Š.; resources and supervision, P.H. and L.K.; writing—review and editing, V.H., L.K. and P.H.; project administration, P.H. and V.H.; funding acquisition, P.H. and K.H. All authors have read and agreed to the published version of the manuscript.

Funding: This research was financially supported by the project CZ.01.1.02/0.0/0.0/15_019/0004399 of Ministry of Industry and Trade of the Czech Republic.

Acknowledgments: Authors also acknowledge the equipment base of research infrastructure IPMinfra (LM2015069) and CEITEC (LQ1601) used during the research activities.

Conflicts of Interest: The authors declare no conflict of interest.

References

1. Lin, H.; Geng, H.P.; Zhang, Y.Y.; Li, H.; Liu, X.Y.; Zhou, X.F.; Yu, L. Fatigue Strength and Life Prediction of a MAR-M247 Nickel-Base Superalloy Gas Turbine Blade with Multiple Carbide Inclusions. *Strength Mater.* **2019**, *51*, 102–112. [CrossRef]
2. Liu, G.; Salvat Cantó, J.; Winwood, S.; Rhodes, K.; Birosca, S. The effects of microstructure and microtexture generated during solidification on deformation micromechanism in IN713C nickel-based superalloy. *Acta Mater.* **2018**, *148*, 391–406. [CrossRef]
3. Liu, Y.; Kang, M.; Wu, Y.; Wang, M.; Li, M.; Yu, J.; Gao, H.; Wang, J. Crack formation and microstructure-sensitive propagation in low cycle fatigue of a polycrystalline nickel-based superalloy with different heat treatments. *Int. J. Fatigue* **2018**, *108*, 79–89. [CrossRef]
4. Lamm, M.; Singer, R.F. The Effect of Casting Conditions on the High-Cycle Fatigue Properties of the Single-Crystal Nickel-Base Superalloy PWA 1483. *Metall. Mater. Trans. A* **2007**, *38*, 1177–1183. [CrossRef]
5. Kunz, L.; Lukáš, P.; Konečná, R.; Fintová, S. Casting defects and high temperature fatigue life of IN 713LC superalloy. *Int. J. Fatigue* **2012**, *41*, 47–51. [CrossRef]
6. Smid, M.; Hornik, V.; Hutar, P.; Hrbacek, K.; Kunz, L. High Cycle Fatigue Damage Mechanisms of MAR-M 247 Superalloy at High Temperatures. *Trans. Indian Inst. Met.* **2016**, *69*, 393–397. [CrossRef]
7. He, L.Z.; Zheng, Q.; Sun, X.F.; Guan, H.R.; Hu, Z.Q.; Tieu, A.K.; Lu, C.; Zhu, H.T. Effect of carbides on the creep properties of a Ni-base superalloy M963. *Mater. Sci. Eng. A* **2005**, *397*, 297–304. [CrossRef]
8. Liu, L.R.; Jin, T.; Zhao, N.R.; Wang, Z.H.; Sun, X.F.; Guan, H.R.; Hu, Z.Q. Effect of carbon addition on the creep properties in a Ni-based single crystal superalloy. *Mater. Sci. Eng. A* **2004**, *385*, 105–112. [CrossRef]
9. Ruttert, B.; Meid, C.; Mujica Roncery, L.; Lopez-Galilea, I.; Bartsch, M.; Theisen, W. Effect of porosity and eutectics on the high-temperature low-cycle fatigue performance of a nickel-base single-crystal superalloy. *Scr. Mater.* **2018**, *155*, 139–143. [CrossRef]
10. Asquith, G.; Pickard, A.C. Fatigue testing of gas turbine components. *High Temp. Technol.* **1988**, *6*, 131–143. [CrossRef]
11. Hu, D. Combined fatigue experiments on full scale turbine components. *Aircr. Eng. Aerosp. Technol.* **2013**, *85*, 4–9. [CrossRef]
12. Kiyak, Y.; Fedelich, B.; May, T.; Pfennig, A. Simulation of crack growth under low cycle fatigue at high temperature in a single crystal superalloy. *Eng. Fract. Mech.* **2008**, *75*, 2418–2443. [CrossRef]
13. Witek, L. Simulation of crack growth in the compressor blade subjected to resonant vibration using hybrid method. *Eng. Fail. Anal.* **2015**, *49*, 57–66. [CrossRef]
14. Chen, L.; Liu, Y.; Xie, L. Power-exponent function model for low-cycle fatigue life prediction and its applications – Part II: Life prediction of turbine blades under creep–fatigue interaction. *Int. J. Fatigue* **2007**, *29*, 10–19. [CrossRef]
15. Lin, B.; Zhao, L.G.; Tong, J.; Christ, H.J. Crystal plasticity modeling of cyclic deformation for a polycrystalline nickel-based superalloy at high temperature. *Mater. Sci. Eng. A* **2010**, *527*, 3581–3587. [CrossRef]
16. Xue, X.; Xu, L. Numerical simulation and prediction of solidification structure and mechanical property of a superalloy turbine blade. *Mater. Sci. Eng. A* **2009**, *499*, 69–73. [CrossRef]
17. Whitesell, H.S.; Li, L.; Overfelt, R.A. Influence of solidification variables on the dendrite arm spacings of Ni-based superalloys. *Metall. Mater. Trans. B* **2000**, *31*, 546–551. [CrossRef]
18. El-Bagoury, N.; Nofal, A. Microstructure of an experimental Ni base superalloy under various casting conditions. *Mater. Sci. Eng. A* **2010**, *527*, 7793–7800. [CrossRef]
19. Holländer, D.; Kulawinski, D.; Weidner, A.; Thiele, M.; Biermann, H.; Gampe, U. Small-scale specimen testing for fatigue life assessment of service-exposed industrial gas turbine blades. *Int. J. Fatigue* **2016**, *92*, 262–271. [CrossRef]
20. Yan, X.J.; Qi, M.J.; Deng, Y.; Chen, X.; Sun, R.J.; Lin, L.S.; Nie, J.X. Investigation on Material's Fatigue Property Variation Among Different Regions of Directional Solidification Turbine Blades-Part II: Fatigue Tests on Bladelike Specimens. *J. Eng. Gas Turbines Power-Trans. ASME* **2014**, *136*, 102503. [CrossRef]
21. Harris, K.; Erickson, G.L.; Schwer, R.E. MAR M 247 derivations—CM 247 LC DS alloy, CMSX® single crystal alloys, properties and performance. In Proceedings of the 1984, Seven Springs Mountain Resort, Champion, PA, USA, 2 October 1984.

22. Janowski, G.M. The Effect of Tantalum on the Structure/Properties of Two Polycrystalline Nickel-Base Superalloys: B-1900+ Hf MAR-M247. Master's Thesis, Michigan Technological University, Houghton, MI, USA, 1985.

23. Kurz, W.; Fisher, D.J. *Fundamentals of Solidification*, 4th ed.; Trans Tech Publications: Aedermannsdorf, Switzerland, 1998.

24. Mujica Roncery, L.; Lopez-Galilea, I.; Ruttert, B.; Huth, S.; Theisen, W. Influence of temperature, pressure, and cooling rate during hot isostatic pressing on the microstructure of an SX Ni-base superalloy. *Mater. Des.* **2016**, *97*, 544–552. [CrossRef]

25. Haocheng, Z.; Anqiang, W.; Zhixun, W.; Zhufeng, Y.; Chengjiang, Z. Effects of Hot Isostatic Pressing (HIP) on Microstructure and Mechanical Properties of K403 Nickel-Based Superalloy. *High Temp. Mater. Process.* **2016**, *35*, 463–471. [CrossRef]

26. Suresh, S. *Fatigue of Materials, 2 ed.*; Cambridge University Press: Cambridge, UK, 1998.

27. Kunz, L.; Lukáš, P.; Konečná, R. High-cycle fatigue of Ni-base superalloy Inconel 713LC. *Int. J. Fatigue* **2010**, *32*, 908–913. [CrossRef]

28. Yan, X.J.; Chen, X.; Sun, R.J.; Deng, Y.; Lin, L.S.; Nie, J.X. Investigation on Material's Fatigue Property Variation Among Different Regions of Directional Solidification Turbine Blades-Part I: Fatigue Tests on Full Scale Blades. *J. Eng. Gas Turbines Power-Trans. ASME* **2014**, *136*, 102502. [CrossRef]

29. Miao, J.; Pollock, T.M.; Wayne Jones, J. Crystallographic fatigue crack initiation in nickel-based superalloy René 88DT at elevated temperature. *Acta Mater.* **2009**, *57*, 5964–5974. [CrossRef]

30. Šulák, I.; Obrtlík, K. AFM, SEM AND TEM study of damage mechanisms in cyclically strained mar-M247 at room temperature and high temperatures. *Theor. Appl. Fract. Mech.* **2020**, *108*, 102606. [CrossRef]

31. Jiang, R.; Bull, D.J.; Evangelou, A.; Harte, A.; Pierron, F.; Sinclair, I.; Preuss, M.; Hu, X.T.; Reed, P.A.S. Strain accumulation and fatigue crack initiation at pores and carbides in a SX superalloy at room temperature. *Int. J. Fatigue* **2018**, *114*, 22–33. [CrossRef]

32. Du, B.; Yang, J.; Cui, C.; Sun, X. Effects of grain size on the high-cycle fatigue behavior of IN792 superalloy. *Mater. Design (1980-2015)* **2015**, *65*, 57–64. [CrossRef]

33. Gao, Y.; Stölken, J.S.; Kumar, M.; Ritchie, R.O. High-cycle fatigue of nickel-base superalloy René 104 (ME3): Interaction of microstructurally small cracks with grain boundaries of known character. *Acta Mater.* **2007**, *55*, 3155–3167. [CrossRef]

34. Liu, G.; Winwood, S.; Rhodes, K.; Birosca, S. The effects of grain size, dendritic structure and crystallographic orientation on fatigue crack propagation in IN713C nickel-based superalloy. *Int. J. Plast.* **2020**, *125*, 150–168. [CrossRef]

35. Leverant, G.R.; Gell, M. The influence of temperature and cyclic frequency on the fatigue fracture of cube oriented nickel-base superalloy single crystals. *Metall. Trans. A* **1975**, *6*, 367. [CrossRef]

36. Bowles, C.Q.; Broek, D. On the formation of fatigue striations. *Int. J. Fract. Mech.* **1972**, *8*, 75–85. [CrossRef]

37. Gunasegaram, D.R.; Farnsworth, D.J.; Nguyen, T.T. Identification of critical factors affecting shrinkage porosity in permanent mold casting using numerical simulations based on design of experiments. *J. Mater. Process. Technol.* **2009**, *209*, 1209–1219. [CrossRef]

38. Flemings, M.C. Solidification processing. *Metall. Trans.* **1974**, *5*, 2121–2134. [CrossRef]

39. Cox, M.; Wickins, M.; Kuang, J.P.; Harding, R.A.; Campbell, J. Effect of top and bottom filling on reliability of investment castings in Al, Fe, and Ni based alloys. *Mater. Sci. Technol.* **2000**, *16*, 1445–1452. [CrossRef]

40. Wu, S.-P.; Li, C.-Y.; Guo, J.-J.; Su, Y.-Q.; Lei, X.-Q.; Fu, H.-Z. Numerical simulation and experimental investigation of two filling methods in vertical centrifugal casting. *Trans. Nonferrous Met. Soc. China* **2006**, *16*, 1035–1040. [CrossRef]

41. Murakami, Y. Material defects as the basis of fatigue design. *Int. J. Fatigue* **2012**, *41*, 2–10. [CrossRef]

42. Tammas-Williams, S.; Withers, P.J.; Todd, I.; Prangnell, P.B. The Influence of Porosity on Fatigue Crack Initiation in Additively Manufactured Titanium Components. *Sci. Rep.* **2017**, *7*, 7308. [CrossRef]

43. Yi, J.Z.; Gao, Y.X.; Lee, P.D.; Flower, H.M.; Lindley, T.C. Scatter in fatigue life due to effects of porosity in cast A356-T6 aluminum-silicon alloys. *Metall. Mater. Trans. A* **2003**, *34*, 1879. [CrossRef]

44. Wang, Q.G.; Apelian, D.; Lados, D.A. Fatigue behavior of A356-T6 aluminum cast alloys. Part I. Effect of casting defects. *J. Light Met.* **2001**, *1*, 73–84. [CrossRef]

45. Horstemeyer, M.F.; Yang, N.; Gall, K.; McDowell, D.; Fan, J.; Gullett, P. High cycle fatigue mechanisms in a cast AM60B magnesium alloy. *Fatigue Fract. Eng. Mater. Struct.* **2002**, *25*, 1045–1056. [CrossRef]
46. Nicoletto, G.; Konečná, R.; Fintova, S. Characterization of microshrinkage casting defects of Al–Si alloys by X-ray computed tomography and metallography. *Int. J. Fatigue* **2012**, *41*, 39–46. [CrossRef]

Publisher's Note: MDPI stays neutral with regard to jurisdictional claims in published maps and institutional affiliations.

MDPI
St. Alban-Anlage 66
4052 Basel
Switzerland
Tel. +41 61 683 77 34
Fax +41 61 302 89 18
www.mdpi.com

Metals Editorial Office
E-mail: metals@mdpi.com
www.mdpi.com/journal/metals

* 9 7 8 3 0 3 6 5 4 6 9 2 6 *